ST291
Images and Information

Coh

Prepared for the Course Team by Stuart Freake

Contents

Study Guide	3
1 Introduction	3
2 How does the illumination affect the diffraction pattern?	4
2.1 Experiment 1 A profound difference	6
2.2 Light from lasers and tungsten lamps	6
2.3 Experiment 2 Producing a narrow parallel monochromatic beam from the tungsten lamp	7
2.4 Experiment 3 Diffraction patterns produced with the pseudo-laser beam	10
2.4.1 The effect of using a broad parallel beam	10
2.5 Effects of the spectrum of the source	10
2.5.1 Experiment 4 Illumination with a broad wavelength spectrum	10
2.5.2 Experiment 5 Light from the laser — a surprising observation	12
2.6 Effects of changing the position and the size of the light source	12
2.6.1 Experiment 6 Changing the source position	12
2.6.2 Experiment 7 Changing the source size	13
2.7 Experiment 8 Broad bandwidth and large source	14
2.8 The convolution theorem revisited	14
2.9 Summary of Section 2	17
3 Light bulbs and lasers	18
3.1 Summary of Section 3	22
4 The addition of waves	23
4.1 The general case	23
4.2 Adding waves with the same temporal frequency	25
4.3 Adding waves with different temporal frequencies	26
4.4 Adding waves from different source points	27
4.5 Summary of Section 4	28
5 Temporal coherence	29
5.1 Introducing coherence	29
5.2 Temporal coherence and bandwidth	30
5.3 Measuring temporal coherence with a double slit	31
5.3.1 The Fourier approach to temporal coherence	31
5.3.2 The Huygens approach to temporal coherence	32
5.4 Quantifying temporal coherence: coherence time and coherence length	37
5.5 The Michelson interferometer	39
5.6 Summary of Section 5	39
6 Spatial coherence	41
6.1 The Fourier approach to spatial coherence	42
6.2 The extent of spatial coherence	44
6.3 The Huygens approach to spatial coherence	46
6.4 Investigating the structure of self-luminous objects	47
6.5 Summary of Section 6	49
7 X-ray diffraction	50
7.1 Summary of Section 7	52
ITQ answers and comments	53
SAQ answers and comments	57
Index of important terms and concepts	60

The ST291 Course Team

Keith Hodgkinson	(Chairman, author)	Ian Every	(Academic Computing Service)
Stuart Freake	(Author)	Richard Sillitto	(External assessor)
Cheryl Newport	(Course Manager)	Michael French	(Consultant)
David Tillotson	(Course editor)	Barrie Jones	(Consultant)
Andrew Millington	(BBC)	Elizabeth Parvin	(Consultant)
Gerald Elliot	(Reader)	Caroline Husher	(Designer)
Ruth Morant	(Reader)	John Taylor	(Graphic Artist)

The current edition of ST291 is an extensive revision of a course first presented in 1977. Several members of the original course team made such a substantial contribution to the design, content, and style of that course that their influence has carried over into this new revised edition.

We should particularly like to acknowledge the contributions of the following people:

Open University: G. S. Bellis, S. M. Freake, K. A. Hodgkinson, B. W. Jones, M. Shott and A. J. Walton. *BBC*: A. B. Jolly, A. J. Millington, E. F. Smith.

The Open University, Walton Hall, Milton Keynes, MK7 6AA.

First published 1992. Reprinted 1994, 1998

Copyright © 1992 The Open University.

All rights reserved; no part of this publication may be reproduced, stored in a retrieval system, or transmitted in any form or by any means, electronic, mechanical, photocopying, recording or otherwise without either the prior written permission of the Publishers or a licence permitting restricted copying issued by the Copyright Licensing Agency, 90 Tottenham Court Road, London, W1P 9HE. This book may not be lent, resold, hired out or otherwise disposed of by way of trade in any form of binding or cover other than that in which it is published, without the prior consent of the Publishers.

Edited, designed and typeset in the United Kingdom by the Open University.

Printed in the Untied Kingdom by Henry Ling Ltd, at the Dorset Press, Dorchester, Dorset.

ISBN 0 7492 5053 4

This text forms part of an Open University Second Level Course. If you would like a copy of *Studying with the Open University*, please write to the Central Enquiry Service, PO Box 200, The Open University, Walton Hall, Milton Keynes, MK7 6YZ. If you have not already enrolled on the Course and would like to buy this or other Open University material, please write to Open University Educational Enterprises Ltd, 12 Cofferidge Close, Stony Stratford, Milton Keynes, MK11 1BY, United Kingdom.

1.3

ST291u5i1.3

ST291 Unit 5

Study Guide

There are four major components in this unit: this printed text, a series of experiments with the Home Kit, an audiovisual sequence, and a video programme.

The experimental work is the major part of Section 2, and I expect that it will take you about two hours to do the experiments and study the associated text. Since some of the experimental observations will be clearer in a darkened room, you may wish to carry out this work in the evening. The only additional materials that you will need to provide for these experiments are Sellotape or Blue-Tack, a pair of scissors, a sharp pointed object for making a 1–2 mm diameter hole in a piece of paper, and a pencil or *non-permanent* felt-tip pen for marking a glass screen.

The audiovisual sequence '*Light bulbs and lasers*' comprises a spoken commentary on Audiocassette 1 and a series of frames printed in Section 3 of this text. This sequence refers to experiments in Section 2, so you should study it *after* doing the experiments, but you could postpone studying it until after later sections if that is more convenient. The commentary is 25 minutes long, and you will need another 20–30 minutes to do the associated ITQs and SAQs.

The video programme '*Michelson interferometer*' develops the concepts introduced in Section 5, and you should therefore view it *after* studying that section. The programme is 25 minutes long, and there are some notes about it in the *Video Notes* booklet.

1 Introduction

In the previous two units, you studied how an object modifies the illumination that falls on it. Those units were particularly concerned with the relationship between the spatial structure of objects and the corresponding diffraction patterns. By a combination of experimental observations and examples in the text, a number of rules were established that describe this relationship. Two of the simplest rules are the symmetry relationship (object and diffraction pattern have similar symmetry) and the inverse relationship (small-scale structure in the object gives rise to large-scale structure in the diffraction pattern and vice versa). These rules, together with the addition rule and the convolution theorem, allow us to relate quite complex objects to their diffraction patterns. So the most important message of Units 3 & 4 is that a diffraction pattern contains coded information about the object, and the various rules are the key to the code. Give me a diffraction pattern, and I can use the 'decoding' rules to tell you quite a lot about the form of the object.

However, there is a caveat that must be applied. These rules relating objects to diffraction patterns apply only to the simple type of illumination that was used in the earlier units. This illumination was a parallel laser beam, which provided *monochromatic* light, in the form of *plane waves*, incident *perpendicular to the object*. If we use a different type of illumination, then we will get a different diffraction pattern. The differences in the diffraction patterns can be quite profound, and they arise because the waves leaving the object are not just characteristic of the object itself, *but also characteristic of the illumination.*

What this means is that the nature of the illumination falling on an object affects the way that information about the object is encoded in the diffraction pattern. However, even though *diffraction patterns* are dependent on the nature of the illumination, it is always possible to form a conventional *image* of the object by using a lens to decode the diffraction pattern. You will be able to demonstrate this for yourself by forming images from diffraction patterns produced with both laser and tungsten lamp illumination in the experiments in a later unit.

You might think that if the coded information contained in the diffraction pattern can so easily be decoded with a lens, then there is little point in exploring the nature of the illumination and precisely how it affects the coding. However, there are many situations in which the nature of the illumination is vitally important, and you will meet some of them later in this Course. Take holography, as one example. Holography is only practicable with laser illumination; you don't make holograms with tungsten lamps. It is therefore important to understand what characteristics of laser light make it suitable for holography, and equally to understand why light from a tungsten lamp is unsuitable for this purpose. Another example of a situation where the nature of the illumination is important is the technique of spatial filtering, which again is the subject of later units. This technique involves selectively modifying parts of a diffraction pattern in order to enhance the image that is subsequently formed. For this to be possible, it is essential that the information about the object is coded in a simple way in the diffraction pattern, and this simple coding is much easier to achieve with laser-type illumination rather than an ordinary tungsten lamp. A third example is the performance of lenses. As you will discover later in the Course, for any lens system there is a maximum spatial frequency that can be imaged, and this maximum frequency depends on the type of illumination being used. Therefore, to characterize the performance of a lens system, it is essential that we know something about the illumination that will be used.

In the previous units, the illumination was restricted to monochromatic plane waves travelling perpendicular to the object, so that we could essentially ignore the details of the illumination and concentrate on the diffracting effect of the object. Now we will return to the real world, where illumination is rarely in the form of parallel laser beams. In this unit, we will examine what happens when the illumination is more complicated. We will explore the effects on the diffraction pattern of using illumination with a range of wavelengths (i.e. different colours), illumination that is not a simple plane wave, and illumination that is not perpendicular to the object.

In the next section, you will be able to see for yourself how the illumination affects the diffraction pattern. Using the Home Experiment Kit, you can compare diffraction patterns produced by illumination from a laser with those produced by illumination from a light bulb. You can then investigate in more detail how the diffraction pattern changes when the range of wavelengths changes, when the size of the source changes, and so on.

Following this experimental investigation, I will discuss in an audiovisual sequence in Section 3 how and why laser illumination differs from the more familiar illumination produced by a light bulb. You will see that the waveforms of these two types of illumination are quite different. Then, in Section 4, I will show how the waveform differences lead to differences in the total intensity observed when several waves of either type are combined.

The following sections will show how important properties of illumination can be described in terms of a concept known as *coherence*. There are two aspects to the coherence of illumination; one is *spatial* coherence, which is related to the spatial geometry of the source, and the other is *temporal* coherence, which is related to the range of wavelengths (or temporal frequencies) in the illumination. I will show how the degree of each type of coherence can be quantified, and how changes to the spatial and/or temporal coherence of the illumination affect the diffraction pattern.

2 How does the illumination affect the diffraction pattern?

In the introduction, I pointed out that the form of the illumination falling on an object can have a profound effect on the diffraction pattern that is produced. I would like you to observe this for yourself by doing a series of experiments using the Home Kit. In the first of these experiments, you will observe differences between the diffraction patterns produced when objects are illuminated by a laser and the corresponding diffraction patterns produced by illumination from a tungsten lamp. Then, in order to understand how these differences arise, you will assemble an optical system that makes the tungsten lamp illumination much more like a laser beam. By modifying this optical system, you will be able to observe not only the effect of changing the range of wavelengths present in the illumination, but also the effect of using illumination made up of plane waves with a range of inclinations.

The observations that you will make can be generalized to account for the forms of diffraction patterns produced by a variety of different types of illumination. For example, knowing the form of the diffraction pattern produced when a monochromatic plane wave beam is incident perpendicular to an object, it is possible to deduce the form of the diffraction pattern produced by any other type of illumination. Or conversely, by inspecting the diffraction pattern of a known (simple) object, it is possible to deduce information about the illumination that is falling on that object.

Now before you embark on the experiments, I want to remind you of the *quantitative* relationship between the spatial frequencies in the object and the positions at which information about those frequencies appears in the diffraction pattern. This relationship was introduced in Units 3 & 4, and is summarized in Figure 1. When an object is illuminated with plane-wave

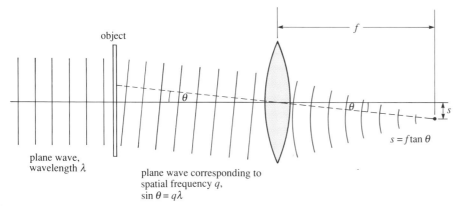

Figure 1 A diffracted plane wave is focused to an off-axis point in the focal plane behind a lens.

monochromatic illumination, with wavelength λ, travelling parallel to the optical axis, the plane waves corresponding to spatial frequency q travel away from the object at angles $\pm \theta$ to the optical axis that are given by the relation

$$\sin \theta = q\lambda. \quad (1)$$

Also, a plane wave travelling at an angle θ to the optical axis is focused by a lens, with focal length f, to a point at a distance s from the optical axis given by

$$s = f \tan \theta. \quad (2)$$

These two equations together allow us to predict where information about a specific spatial frequency q will appear in the diffraction pattern (assuming wavelength λ and focal length f are known), or, conversely, to determine what spatial frequencies are present in the object by measuring values of s (again assuming λ and f are known).

Such calculations are much simpler when the angle θ is small, because then $\sin \theta \approx \tan \theta$, and equations 1 and 2 can be combined to produce the relation

$$s = fq\lambda. \quad (3)$$

But always bear in mind that this result is only accurate *if the angle θ is small*. How small is small? Well, it depends on the accuracy you require from the calculation. For example, if you are interested in 1% accuracy, then equation 3 is good enough provided that $\theta < 8\frac{1}{2}°$, or $\tan \theta < 0.15$. So if s/f is less than 0.15, results obtained by using equation 3 will be accurate to better than 1%. For $\theta > 8\frac{1}{2}°$, you would have to use equations 1 and 2 to get an accuracy of better than 1%. Table 1 shows that the accuracy obtainable from equation 3 becomes very much poorer as the values of θ and of $\tan \theta$ ($= s/f$) increase.

Table 1 Accuracy of the equation $s = fq\lambda$ for various values of the diffraction angle θ

θ	$\tan \theta$	accuracy of $s = fq\lambda$
1°	0.0174	0.015%
3°	0.0524	0.14%
10°	0.176	1.5%
15°	0.268	3.5%
20°	0.364	6.4%
25°	0.466	10.3%

Study note

The experiments associated with this unit are concentrated into Sections 2.1–2.7. You will require about 2 hours to do these experiments and study the associated text.

2.1 Experiment 1 A profound difference

In this first experiment, you will see the contrast between diffraction patterns that are produced with a laser beam and those produced with light from a tungsten lamp.

- Set up the optical bench, with the laser at one end and the clear tungsten lamp at the other, as shown in Figure 2.

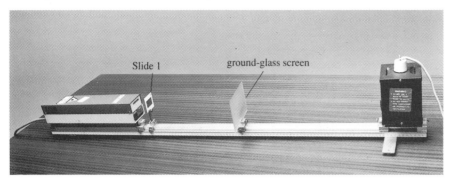

Figure 2 Home Kit set up for Experiment 1.

- Mount the large ground-glass screen in a saddle mid-way between the laser and the lamp. *Take care not to crack the glass*: you should place a piece of thin card between the glass and the saddle screw before tightening it, and you should only tighten the screw *very gently*.

I suggest that you investigate the diffraction patterns produced by a few of the 'objects' on Slide 1. The layout of the objects on this slide is shown in Figure 3. Objects 4, 10 and 13 are particularly suitable for this investigation since the diffraction patterns produced with the laser beam show quite detailed structure (as you have observed already in Units 3 & 4).

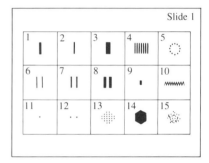

Figure 3 Objects on Slide 1.

- Mount Slide 1 in a post and saddle close to the laser, and adjust its position so that the laser beam passes through the first object that you wish to use.

- Inspect the diffraction pattern on the ground-glass screen, adjusting the slide position if necessary to make sure that the beam passes through the centre of the object. Make a sketch of the diffraction pattern in the margin, or make a few notes about its appearance.

- Now switch off the laser and switch on the tungsten lamp. Because the tungsten lamp produces a broad beam, it is necessary to make an aperture so that light can only reach the object of interest. Cut a small hole, three or four millimetres square, in a piece of the black paper provided in the Kit, and stick the paper to the frame of Slide 1 with Blue-Tack or Sellotape so that light can only pass through the first object that you are investigating. Place Slide 1 in front of the opening in the lamp-housing, and, if necessary, adjust its height and lateral position so that the transmitted light falls near the centre of the screen. Can you see any similarity between this diffraction pattern and the one you observed previously with the laser beam, or are they completely different? Record your observations in the margin.

- Repeat the comparison of diffraction patterns produced with the laser and with the tungsten lamp using several other objects on Slide 1.

Your observations should have convinced you that the form of the diffraction pattern is strongly dependent on the illumination used. In particular, you will now agree, no doubt, that the parallel laser beam produces a more detailed, sharper diffraction pattern — ample justification for its use in Units 3 & 4.

2.2 Light from lasers and tungsten lamps

Laser illumination differs in several respects from the illumination produced by a tungsten lamp. Two of the differences are immediately obvious to the eye, and these differences account for the fact that the diffraction patterns produced with the two types of illumination bear little resemblance to one another.

☐ What are the two most obvious differences between the light from the laser and the light from the tungsten lamp?

■ I think that the most obvious differences are (i) the directionality of the illumination and (ii) the colour of the illumination, which is determined by the range of wavelengths present.

Lasers are noted for their narrow, almost parallel, beams, whereas the filament of a tungsten lamp emits light in all directions. One rather impressive example of the high directionality of laser beams was provided by experiments associated with the Apollo 11 and Apollo 14 lunar missions. Astronauts from each mission placed reflectors on the surface of the Moon, and pulses of laser light were beamed at these reflectors from the Earth. By the time that the laser light had travelled the 3.84×10^5 km distance between the Earth and the Moon, the beam had spread to a diameter of only about 3 km, which represents an angular divergence of $3 \text{ km}/3.84 \times 10^5 \text{ km} = 8 \times 10^{-6}$ radians. This tiny divergence meant that the intensity (i.e. watts per square metre) of the illumination reaching the reflector was larger by a factor of about $(3.84 \times 10^5/3)^2 \approx 10^{10}$ than would have been the case if the light had radiated equally in all directions. By making precise measurements of the time delay before a reflected laser pulse returned (about $2\frac{1}{2}$ seconds), it was possible to determine the distance to the Moon to an accuracy of about 15 cm, which is one part in 10^9.

The second important difference between light from a laser and light from a tungsten lamp is in their **spectral bandwidth**, that is in the range of wavelengths (or frequencies) present in the light. The illumination from a tungsten lamp contains wavelengths throughout the visible range, from violet light with wavelength about 400 nm to red light with wavelength about 700 nm. (It also contains infrared radiation, which has wavelengths longer than 700 nm.) In contrast, the beam from your helium–neon laser is red and contains only an extremely narrow range of wavelengths. The average wavelength is 632.8 nm and the wavelength range is only about 10^{-2} nm. The contrast between the two types of illumination can be displayed by plotting the wavelength **spectrum** for each, that is a graph of intensity versus wavelength, and this has been done in Figure 4. The spectral bandwidth of the laser illumination (10^{-2} nm) is approximately 10^{-4} times the range of wavelengths in the visible radiation from the tungsten lamp, and so, on the scale of the graph in Figure 4, the laser spectrum is simply represented by a narrow line.

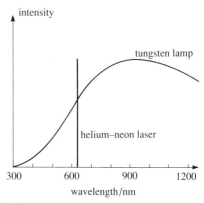

Figure 4 The spectra of illumination from a tungsten lamp and from a helium–neon laser.

The directionality of the illumination falling on an object has a distinctly different effect on the diffraction pattern from the effect caused by the spectral bandwidth, and you will be able to investigate these effects in the experiments that follow. In these experiments, you will first convert the illumination from the tungsten lamp into a narrow parallel monochromatic beam — a simulated laser beam if you like — and you will then check that the diffraction pattern produced by this beam is similar to that produced by the laser. You will subsequently modify the bandwidth and directionality of the illumination independently to observe the effects of these changes on the diffraction pattern.

2.3 Experiment 2 Producing a narrow parallel monochromatic beam from the tungsten lamp

The experiments in Units 1 & 2 will have given you some experience of setting up optical systems using components from the Home Kit. In particular, they will have made you aware of the importance of aligning carefully each component as you build up an optical system, and you should bear this in mind as you assemble the system for producing a narrow parallel monochromatic beam. The steps in the assembly are outlined below, and the layout of the components is shown in Figure 5.

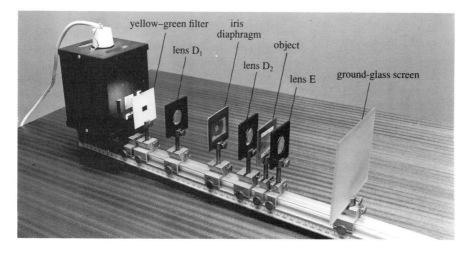

Figure 5 The optical system used for Experiment 2.

- Mount the box containing the tungsten filament lamp in two saddles at one end of the bench. Make sure that the filament is centred over the bench.

In order to control the spatial extent of the light beam, it is necessary to use a **diaphragm** (an opaque disc with a hole in the centre) to restrict the region of the filament that contributes to the beam. Obviously you cannot put such a diaphragm directly in front of the filament, since it is inside a glass bulb. Instead you must use a lens to produce an image of the filament in a more accessible place, and then use the diaphragm to select light from part of this image.

- Use an optical post and a saddle to mount one of the type D lenses ($f \approx 50$ mm) about 100 mm from the front of the light box. I will refer to this lens as D_1 to distinguish it from the second type D lens that you will shortly be adding to the optical system, and which I will refer to as lens D_2.

- Adjust the height of lens D_1 so that the centre of the light beam remains at a constant height for about 30 cm from the lens. Remember, the easiest way to check this is to mount the white card with a printed scale (provided in the Kit) in a short post in a saddle on the optical bench, and to move the card back and forth along the bench while observing the illuminated area on the card. When you have the correct height for lens D_1, adjust its sideways position so that the centre of the light beam remains above the centre of the optical bench.

With lens D_1 correctly aligned, the image of the filament will lie on the optical axis, over the bench, as shown in Figure 6(a); if the lens is misaligned, the image will be offset from the axis, as shown in Figure 6(b).

The next step is to introduce a diaphragm to control the region of the image of the filament that contributes to the illumination. The iris diaphragm from the Kit, which has an aperture that can be adjusted from about 1 mm diameter up to about 20 mm, can be used for this purpose.

- Use Blue-tack or Sellotape to fix the iris diaphragm to a blank 35 mm slide holder, taking care not to obstruct the adjusting lever and not to gum up the delicate mechanism. Mount the slide plus diaphragm in an optical post and saddle on the bench, set the aperture to its minimum diameter, and position the diaphragm so that a sharp image of the filament can be seen on the diaphragm. Adjust the height and sideways position of the diaphragm so that its centre lies at the centre of the filament image.

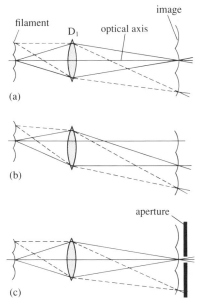

Figure 6 (a) Correct alignment of lens D_1: filament, lens D_1 and filament image are all centred on an axis parallel to the optical bench. (b) Misalignment of lens D_1 results in an inclined light beam and an off-axis filament image. (c) Use of an aperture to select part of the filament image.

You now have a beam diverging from a small area of the filament, as shown in Figure 6(c), and you need to convert this to a parallel beam. This can be done with a second convex lens.

☐ If this second lens has a focal length f, how far must it be placed from the diaphragm if it is to produce a parallel beam?

■ The separation required is equal to the focal length f, as shown in Figure 7(a). Remember that the focal length is the distance from the lens of the point at which a parallel beam comes to a focus, or conversely, the distance from the lens of a point source (i.e. the small aperture in the diaphragm) if the lens is to produce a parallel beam.

If the separation is not equal to the focal length, then the beam will either continue to diverge (Figure 7(b)), or will first converge and then diverge (Figure 7(c)).

- Mount the second type D lens ($f \approx 50$ mm) on the optical bench, and adjust its distance from the diaphragm until it produces an approximately parallel beam. Check for parallelism by moving the white card back and forth along the bench. Since the diaphragm lets through a short length of the filament's image rather than just a point of the image, the beam will tend to diverge in the horizontal plane: each point on the filament produces a parallel beam at a slightly different inclination when the lens is at the correct distance from the diaphragm.

- Adjust the height and sideways position of lens D_2 so that the beam is horizontal and centred over the bench.

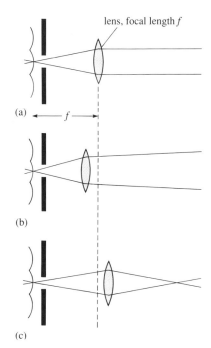

Figure 7 (a) Producing a parallel beam: the lens should be at a distance f from the filament image. (b) A diverging beam is produced when the lens is too close. (c) When the lens is too far from the filament image, the beam converges and then diverges.

The parallel beam that you have produced will be used to illuminate various objects, and an additional lens will be used to form diffraction patterns of these objects in its back focal plane (i.e. the focal plane behind the lens). These diffraction patterns can be viewed on a ground-glass screen placed in the back focal plane. It is simpler to set up this lens and screen without an object in place, and that is the next step to tackle.

- Place a post in a *narrow* saddle immediately after lens D_2, and then mount lens E in a post and another *narrow* saddle immediately after this. (Using narrow saddles here means that lenses D_2 and E are as close as possible.)
- Adjust the height and sideways position of lens E so that the beam it produces is horizontal and centred over the bench.
- Mount the large ground-glass screen in a saddle on the bench, being careful as before not to break the glass. Move the screen back and forth to find the position at which a sharp image of the filament is produced, and clamp the saddle in that position.

Now that you have reached this stage, you have completed the trickiest part of the setting up procedure. You just need to add two more components to your optical system to make the illumination more laser-like: a colour filter to make the beam reasonably monochromatic, and a small aperture to restrict the diameter of the parallel beam to about 2 mm.

- Mount the yellow–green filter in a post and saddle immediately in front of the light box.
- Make a hole with a pointed object, about 2 mm in diameter, in the centre of a 5 cm square of the black paper that is provided in the Kit. Stick this black paper to the mounting of lens D_2, with the hole at the centre of the lens, to restrict the diameter of the parallel beam.

Before you use the narrow parallel monochromatic beam that you have produced, I will review very briefly the function of each of the optical components on your bench. Working from the light source to the screen, i.e from left to right in Figure 5 and the ray diagram in Figure 8,

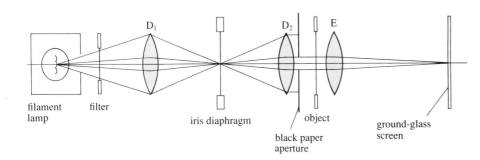

Figure 8 A ray diagram for the apparatus shown in Figure 5. The light from the tungsten lamp is converted to a narrow parallel monochromatic beam, which is diffracted by an object, and a diffraction pattern observed on the ground-glass screen.

  the tungsten lamp is an omnidirectional source of white light;

  the yellow–green filter only transmits light within a narrow band of wavelengths in the yellow–green region of the spectrum;

  lens D_1 produces an image of the filament in an accessible region;

  the iris diaphragm, which is in the plane where the filament image is produced, controls the part of the filament contributing to the illumination;

  lens D_2 produces a parallel beam;

  the small aperture in the black paper transmits a narrow parallel beam, diameter about 2 mm;

  the object (yet to be inserted!) will sit in this parallel beam;

  lens E will focus a diffraction pattern of the light transmitted by the object onto the **ground-glass screen**.

The purpose of setting up this optical system is to investigate how the diffraction pattern from an object depends on the nature of the illumination, and you can now do this in three stages. First, you can check that the beam you have produced, which I will call a pseudo-laser beam, gives similar diffraction patterns to those given by the laser. You can then look in turn at the effect of increasing the spectral bandwidth of the illumination and at the effect of increasing the source size.

2.4 Experiment 3 Diffraction patterns produced with the pseudo-laser beam

The effects that the following experiments should demonstrate will be most readily apparent if you use an object that has a transmittance profile that gives a simple and familiar diffraction pattern. For this reason, I suggest that you use the 300 lines per millimetre grating. This grating has a transmittance with spatial frequency components at $0, \pm 300$ cycles mm^{-1}, ± 600 cycles mm^{-1}, ± 900 cycles mm^{-1}, etc., and so the diffraction pattern formed with laser illumination would be a set of equally spaced spots. (Note that this grating is *not* made up from equal-width dark and clear strips: as you saw in Units 3 & 4, such a grating only has odd harmonics in its Fourier transform, i.e. the components $q_2 = \pm 600$ cycles mm^{-1}, $q_4 = \pm 1200$ cycles mm^{-1}, etc. would not be present.)

- Mount the 300 lines per millimetre grating between lens D_2 and lens E. The lines of the grating should be vertical, and this will be the case if the *300 lines/mm* label is horizontal.

- Observe the diffraction pattern by viewing it through the ground-glass screen. (You will need to darken the room to see the pattern clearly.) Sketch the diffraction pattern in the margin. Measure the separation of the spots on the glass screen: you will find it easiest to do this by marking the positions of the spots on the screen with a pencil or *non-permanent* felt-tip pen, and then removing the screen to make the measurements.

    ITQ 1 (a) Is the diffraction pattern produced by your pseudo-laser similar to the pattern that would be produced by the helium–neon laser?

    (b) Use your measured values of the spot separations to determine the value of the wavelength of the yellow–green light transmitted by the colour filter.

2.4.1 The effect of using a broad parallel beam

The observation and measurement that you have just made used a narrow beam, about 2 mm diameter, to illuminate the grating. This was simply to make the beam similar to the unexpanded laser beam.

   ITQ 2 What changes to the diffraction pattern would you expect to observe if you increased the diameter of the beam illuminating the grating by increasing the diameter of the aperture in the black paper mounted on lens D_2? *Before* referring to the answer at the back of the unit, check your answer experimentally by observing the effect on the diffraction pattern of removing the black paper from lens D_2.

2.5 Effects of the spectrum of the source

For your experiments so far, you have used a coloured filter to limit the spread of wavelength to a narrow band in the yellow–green region of the visible spectrum. The purpose of the next two experiments is to investigate how changes in the wavelengths present in the illumination affect the diffraction pattern.

2.5.1 Experiment 4 Illumination with a broad wavelength spectrum

The experimental components should be set up in the same way as for the previous experiment (see Figures 5 and 8). The 300 lines per millimetre grating should be in the object position, and the colour filter should still be in place, but the black paper diaphragm should no longer be on lens D_2.

The spectral effects are most easily seen if a small source is used.

- Set the aperture in the iris diaphragm to its minimum diameter.
- Check that there are pencil marks on the ground-glass screen just below each of the spots in the diffraction pattern of the grating.
- Now remove the yellow–green filter, and look carefully at the diffraction pattern on the screen. Make a rough sketch in the margin of what you observe, and use your observations to answer the following ITQ.

ITQ 3 (a) Does removal of the filter lead to any overall shift in the diffraction pattern? Is there any shift in the yellow–green part of the pattern?

(b) Explain why the red parts of the diffraction pattern are further from the centre than the corresponding blue parts.

(c) By making appropriate measurements on the diffraction pattern, estimate the wavelengths of red and blue light. (Record these values in the margin: you will need them for SAQ 8.)

(d) Why is the central spot in the diffraction pattern white?

I suggested that you use the 300 lines per millimetre grating again for this experiment because its simple diffraction pattern makes it relatively easy to see the effects of having a range of wavelengths present in the illumination. But the effects are exactly the same, however complicated the diffraction pattern may be. Each different wavelength produces a similar diffraction pattern, but the patterns have different colours and the scale of each pattern is proportional to the wavelength. The total pattern is simply the *sum* of the patterns produced by the individual wavelengths. Figure 9 shows how this works when there are just two wavelengths present in the illumination. For clarity, the diffraction patterns produced by the different wavelengths are represented by open and filled circles, and in each example the wavelength that produced the open circles is 25 per cent larger than the wavelength that produced the closed circles. Note that the zero spatial frequency components always coincide at the centre of each pattern.

The experiment in this section illustrated how the different wavelengths, or the different colours if you like, are spread out in diffraction patterns. From knowledge of the Fourier transform of the transmittance of an object, together with knowledge of the wavelengths present in the illumination, it is straightforward to predict the form of the diffraction pattern that will be produced. But the converse procedure is of much greater practical importance: we can use observations and/or measurements of the form of a diffraction pattern to deduce information about the wavelength spectrum of the illumination, and this is exactly what you did when answering ITQ 3(c)

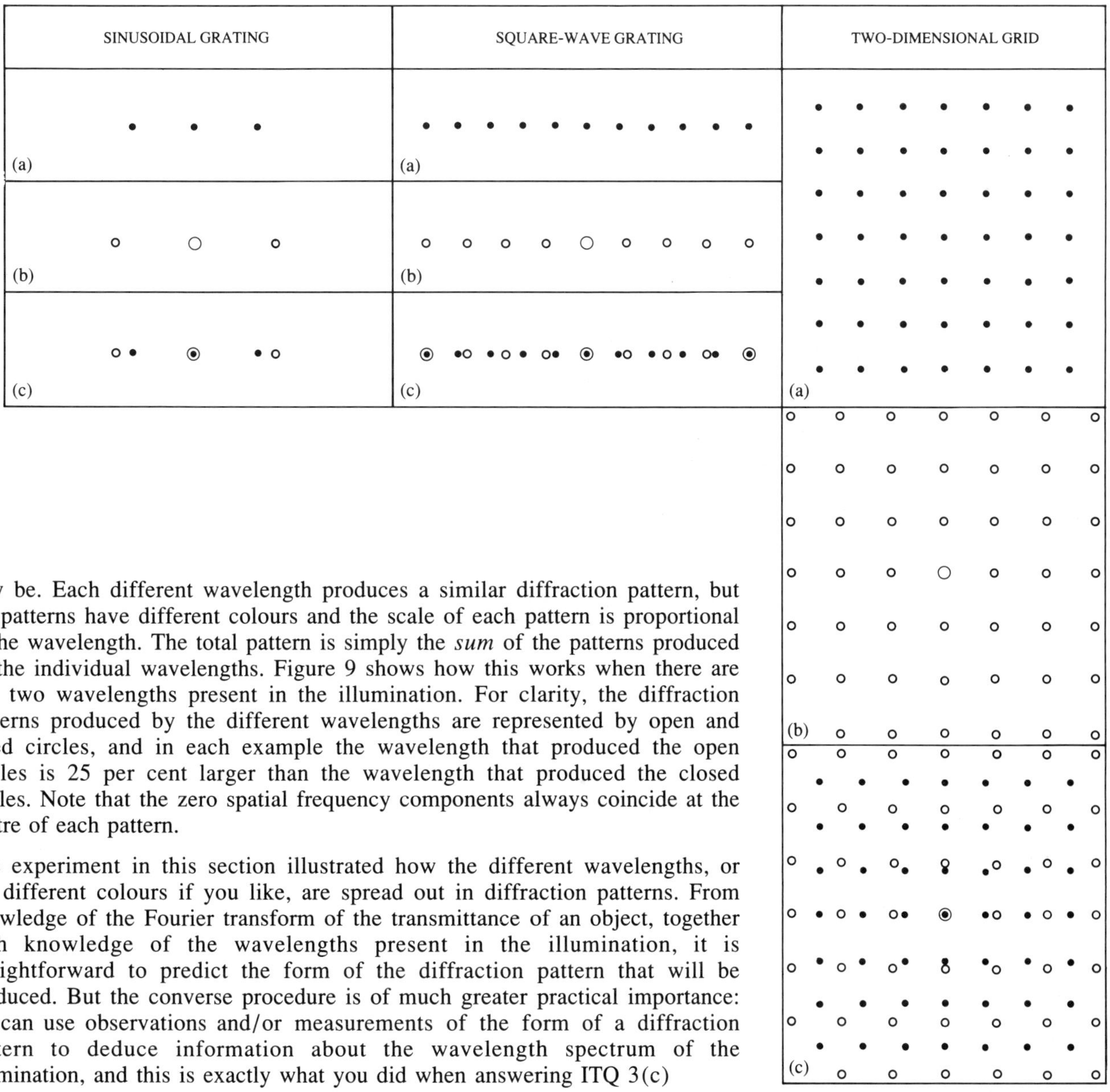

Figure 9 Diffraction patterns produced by three objects when illuminated by parallel beams with different wavelengths. The patterns in row (a) are obtained with a monochromatic beam, wavelength λ_1. Row (b) patterns are obtained with wavelength λ_2, where $\lambda_2 = 1.25\lambda_1$, and row (c) patterns are obtained with a beam containing *both* wavelengths, λ_1 and λ_2.

2.5.2 Experiment 5 Light from the laser — a surprising observation

In this experiment, you will investigate the spectrum of light from the laser by observing a diffraction pattern produced by it. You may think that this will not provide any new information; after all, you already know that the laser produces a parallel beam, and that the light is monochromatic, with a wavelength of 632.8 nm and a very narrow bandwidth of about 10^{-2} nm. If this beam were used to illuminate the 300 cycles per millimetre grating, you would see a row of (approximately) equally spaced red spots, and this would not add to your knowledge about the laser. But is *all* of the light from a laser concentrated into a very narrow wavelength range? And does *all* of the light emerge from the end of the laser in a narrow parallel beam?

You may already have noticed that some orangy-pink light is emitted through the top of the Home Kit laser tube; it can be seen through the two holes in the top of the metal casing of the laser, and it is this light that you should use in this experiment.

- Put the little round magnet from the Home Kit on top of the laser so that it encircles one of the holes in the casing. (Choose the hole through which most light seems to be coming.) Put the two razor blades onto the magnet so that they form a narrow slit — the narrower the better, but make sure that some light is getting through.
- Now hold the 300 lines per millimetre grating close to your eye, with the grating lines parallel to the slit created by the razor blades, and look through the grating at the light emerging from the slit.
- Do not look *straight* at the slit, but look slightly to one side.
- Sketch in the margin what you can see, and then answer the following ITQ.

   ITQ 4 Is the spectrum of the light from the top of the laser tube more like the spectrum of the laser beam or more like the spectrum of the tungsten lamp?

We will take up the question of why the light from the top of the laser tube differs from the light in the laser beam, and why both differ from the light from the tungsten lamp, in Section 3.

2.6 Effects of changing the position and the size of the light source

In the previous two experiments, you investigated how changes to the wavelength spectrum of the illumination affected the diffraction pattern. Now you will look at the effects of changing source position and source size. These effects will be easier to see with monochromatic light, so the yellow–green filter will need to be used again.

2.6.1 Experiment 6 Changing the source position

In this experiment you will move the iris diaphragm so that it transmits different parts of the image of the tungsten filament, and will observe the effect that this has on the diffraction pattern of the 300 lines per millimetre grating.

- First, replace the yellow–green filter and the 300 lines per millimetre grating in the optical system in the positions shown in Figure 5.
- Check that the aperture in the iris diaphragm is set to its minimum diameter and is located at the centre of the filament image.
- Make sure that the positions of the diffraction spots are marked on the ground-glass screen so that you will be able to observe any changes that take place when the iris diaphragm is moved.
- Now move the iris diaphragm slightly to one side so that the aperture transmits an off-centre part of the filament image. Observe the diffraction pattern: how has it changed?
- Move the aperture to other parts of the filament image, and observe how the diffraction pattern changes. Also, compare the diffraction patterns on the screen when the grating is in position and when the grating is removed.

ITQ 5 Figure 10 shows the path taken by light from the centre of the filament image to the zero frequency spot in the diffraction pattern in the back focal plane of lens E.

(a) Suppose that the aperture is moved to point P. Sketch on this diagram the path taken by light from point P on the filament as it travels to the grating. Sketch also the path of the zero frequency component of the illumination from P as it travels between the grating and the back focal plane of lens E.

(b) In terms of the angle δ marked on Figure 10, (i) what is the inclination to the optical axis of the illumination falling on the grating from point P on the filament, and (ii) what is the distance of the corresponding zero frequency spot from the optical axis in the back focal plane of lens E, which has focal length f_E?

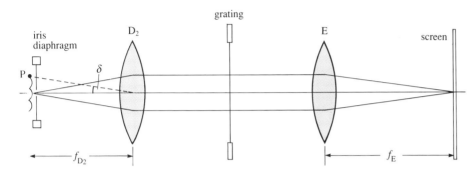

Figure 10 For use with ITQ 5.

The off-axis source produces a diffraction pattern with its centre displaced from the optical axis. As long as the angular displacement of the source point from the axis is small, the displacement of the diffraction pattern's centre is proportional to the displacement of the source. In addition, in this small angle situation, the spacing of the spots in the diffraction pattern remains the same when the pattern shifts; you should have noticed in your experiment that the whole diffraction pattern was displaced when the source point was moved.

2.6.2 Experiment 7 Changing the source size

In the previous experiment, you moved the aperture to select light from different points on the filament, but kept the aperture size constant. You should now investigate what happens to the diffraction pattern when you increase the size of the aperture. Before you do the experiment, try to predict what will happen to the diffraction pattern; your observations in the previous experiment should be useful in making this prediction.

- Before opening the aperture, position it once again at the centre of the filament image produced by lens D_1, and make sure that the positions of the diffraction spots are marked on the glass screen.

- Now increase the size of the aperture while observing the diffraction pattern. Make a note in the margin of the changes that you observe. Compare the diffraction pattern with the pattern on the screen when the grating is removed.

ITQ 6 How did the diffraction pattern change when you opened the aperture? Can you relate this change to the observations you made when you changed the *position* of the aperture?

At the end of this unit, I will show that there is a quantitative relationship between the size of the source and the blurring of the diffraction pattern. So, while you have the optical system set up, you should obtain some data that can be used to confirm this relationship.

- Adjust the size of the aperture so that the filament images produced on the glass screen by the straight-through beam and the 300 cycles mm^{-1} beams touch, but do not overlap.

- Measure the distance between the aperture and lens D_2, and measure the diameter of the aperture. Record your results in the margin: you will need them for SAQ 11 in Section 6.5.

2.7 Experiment 8 Broad bandwidth *and* large source

You have investigated separately how a diffraction pattern is affected by the wavelength spectrum of the illumination and by the size of the source producing the illumination. What happens to the pattern, though, when we use illumination that not only has a broad bandwidth but also comes from a large source? In the following experiment you can investigate this for yourself.

- Remove the yellow–green filter.
- Observe the diffraction pattern for a range of different settings of the aperture diameter.
- Set the aperture diameter to about 5 mm. Rotate the diffraction grating so that the grating lines are horizontal rather than vertical, and observe the diffraction pattern.

   ITQ 7 Describe the diffraction patterns that you observed, and relate them to the patterns observed in the earlier experiments.

2.8 The convolution theorem revisited

The observations that you have made of the effects on a diffraction pattern of the *position and size* of the source of illumination can be very neatly summarized in terms of the convolution theorem that you met in Units 3 & 4, and this will be done in this section. I will assume initially that the illumination is *monochromatic* — so I will be concerned with interpreting Experiments 6 and 7. At the end of the section I will indicate how the analysis applies when a range of wavelengths is present.

A wide variety of objects and diffraction patterns were discussed in Units 3 & 4. Fortunately, it is not necessary to remember all of these object–diffraction pattern pairings because there are a number of simple rules that allow us to deduce complicated pairings from a knowledge of a handful of simple pairings. The most basic rules concern the scaling, orientation and symmetry of object and diffraction pattern. However, you met another rule that is useful when the object transmittance can be regarded as the product of two functions, and this was the convolution theorem. Figure 11 should remind you how the **convolution theorem** is used to find the Fourier transform of a restricted-width square-wave grating.

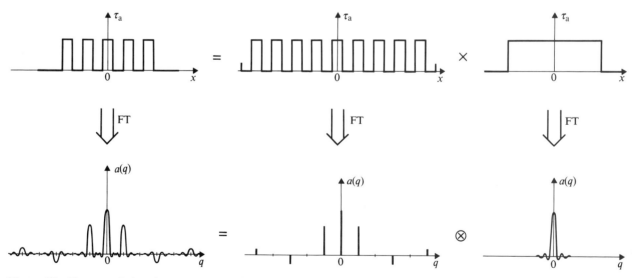

Figure 11 The convolution theorem accounts for the form of the Fourier transform of a restricted square-wave grating. The restricted square-wave grating (top left) is equivalent to the *product* of an infinite square-wave grating (top centre) and a top-hat function (top right). The Fourier transform of the restricted grating (bottom left) is the *convolution* of the Fourier transform of the infinite grating (bottom centre) and the Fourier transform of the top-hat function (bottom right). Bearing in mind that we can think of convolution as a *dealing* operation, the Fourier transform of the restricted grating is obtained by dealing the sinc function, which is the Fourier transform of the top-hat, to each of the delta functions in the Fourier transform of the grating. Of course, the observed intensity in the diffraction pattern for the restricted grating is then found by squaring the Fourier transform.

In Units 3 & 4, you saw many other examples of the application of this convolution theorem. You also saw that it could be represented symbolically as follows:

if $\quad A = B \times C$
then $\quad \mathrm{FT}(A) = \mathrm{FT}(B) \otimes \mathrm{FT}(C) \quad$ } convolution theorem $\hfill (4)$

In the example in Figure 11, A, B and C represent the transmittances of a restricted square-wave grating, an infinite grating and a top-hat function, respectively. $\mathrm{FT}(A)$ represents the Fourier transform of A; in our example, this is the Fourier transform of the restricted grating's transmittance.

Now, how does the convolution theorem help us to interpret the effects on the diffraction pattern of different types of illumination of the object? Well, remember that the lens produces a Fourier transform of the field immediately behind the object. The key point is that the field behind the object is the *product* of the illuminating field and the grating transmittance. In symbols, we can write

$$E_{\mathrm{out}} = E_{\mathrm{in}} \times \tau_{\mathrm{a}}$$

where E_{in} and E_{out} are the fields in front of and behind the object transparency. The field in the back focal plane of the lens is $\mathrm{FT}(E_{\mathrm{out}})$, the Fourier transform of the output field.

☐ Use the convolution theorem to express the field in the back focal plane in terms of the input field E_{in} and the transmittance τ_{a}.

■ Since $\quad E_{\mathrm{out}} = E_{\mathrm{in}} \times \tau_{\mathrm{a}}$
then $\quad \mathrm{FT}(E_{\mathrm{out}}) = \mathrm{FT}(E_{\mathrm{in}}) \otimes \mathrm{FT}(\tau_{\mathrm{a}}) \quad$ } convolution theorem $\hfill (5)$

The field in the back focal plane is $\mathrm{FT}(E_{\mathrm{out}})$, so this field is the convolution of the Fourier transform of the input field with the Fourier transform of the object transmittance.

You may wonder why we have ignored this 'convolution of the Fourier transform of the input field' in previous units. The reason is simple. In previous units, we always used monochromatic plane wave illumination travelling parallel to the optical axis. This produces illumination E_{in} that is constant in space across the object. In this case, the spatial variation of E_{out} is determined only by τ_{a}, and this is illustrated in Figure 12(a) for the familiar sinusoidal transmittance. Since the constant E_{in} makes no contribution to the variation of E_{out}, it can have no influence on the diffraction pattern, and this is why we ignored the input field.

We can arrive at exactly the same conclusion by using an alternative argument that invokes the convolution theorem. The Fourier transform of a constant-in-space field is a delta function located at the origin in the diffraction pattern, and convolution of any function with a delta function at the origin leaves the function unchanged. This is represented graphically in Figure 12(b) for the Fourier transform of the sinusoidal grating.

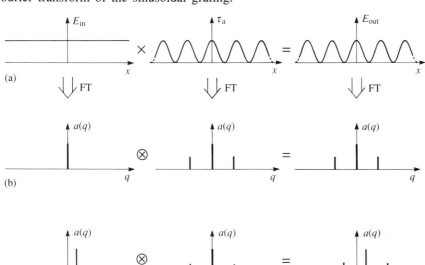

Figure 12 (a) When E_{in} is constant in space, the output field E_{out} has the same form as the transmittance τ_{a}. (b) Convolution of the Fourier transform of a constant-in-space field with the Fourier transform of a sinusoidal transmittance leaves the latter unchanged. (c) The transform of an inclined parallel beam is an off-axis delta function, and convolution now leads to a displacement of the transform of the sinusoidal transmittance.

Now consider what happens when the parallel beam is inclined to the optical axis. An inclined parallel beam will be focused by the lens to a point off the axis in the back focal plane (Figure 54). In terms of Fourier transforms, this is equivalent to saying that the transform of an inclined plane wave is a delta function that is offset from the origin. So when the 'three-spike' pattern (that is, the Fourier transform of the sinusoidal grating) is dealt out to this offset delta function, the three-spike function itself is offset from the axis by the same amount as the delta function. This is shown graphically in Figure 12(c). The effect of inclining the plane wave beam is to shift the diffraction pattern: it is no longer centred on the optical axis.

> **ITQ 8** Suppose that the sinusoidal grating is illuminated simultaneously by two monochromatic plane waves, with identical wavelengths, but at different angles of inclination. One is at the same inclination as in the example in Figure 12(c), and the second at double this inclination.
>
> (a) Sketch three graphs, analogous to those in Figure 12(c), to illustrate how the Fourier transform of the output field from the grating is obtained by a convolution operation.
>
> (b) Draw sketches to show the diffraction pattern in the back focal plane of the lens (i) when the object grating is removed, and (ii) when the object grating is in place.

The application of the convolution theorem to an extended source (like the tungsten filament) is straightforward, since an extended source can be thought of as many point sources packed close together. Each of these point sources, in conjunction with collimating lens D_2, produces a parallel beam, but the inclinations of the beams cover a continuous range of angles. The Fourier transform of this illumination is therefore a set of delta functions, spread out continuously along a line, as shown in the graph on the left of Figure 13. This continuous distribution of delta functions is essentially an image of the source, with each point on the source producing a point in the image in the back focal plane. The convolution of this continuous distribution of delta functions with the three-delta-function transform of the grating transmittance produces the Fourier transform shown in the graph on the right of Figure 13. The diffraction pattern shown below the graph corresponds to the case where the source is an extended filament, as in your tungsten lamp.

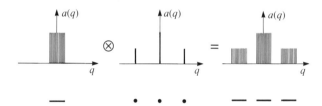

Figure 13 The convolution theorem applied to illumination from an extended source.

A very similar analysis applies when the source is two dimensional. It is difficult to represent this graphically, but the relationship between the diffraction patterns can easily be shown. Figure 14 illustrates this for the sinusoidal grating illuminated with a source that has the shape of the letter P. The diffraction pattern is a convolution of the diffraction pattern of the grating transmittance with the P-shaped pattern that would be produced by the illumination in the absence of grating.

P ⊗ • • • = P P P

Figure 14 A two-dimensional convolution: the P-shaped intensity pattern produced by the source is dealt out to the three delta functions that correspond to the Fourier transform of the sinusoidal grating.

Finally, we need to consider the most general case, where the illumination contains a range of different wavelengths as well as being produced by an extended source. Experiment 4 showed the effect of different wavelengths with a point source of illumination. Essentially, for a given object, all wavelengths produce diffraction patterns with the same general form, but each pattern has the colour appropriate to that wavelength, and the spatial scale of the pattern is proportional to the wavelength. The total diffraction pattern is simply the *sum* of the patterns due to the different wavelengths present in the illumination. Note carefully that because the diffraction patterns corresponding to different wavelengths are different in colour and scale, we cannot use the convolution theorem to relate the pattern produced by broad bandwidth illumination to the pattern produced by monochromatic illumination.

Frame 7: Three processes involving atoms and photons

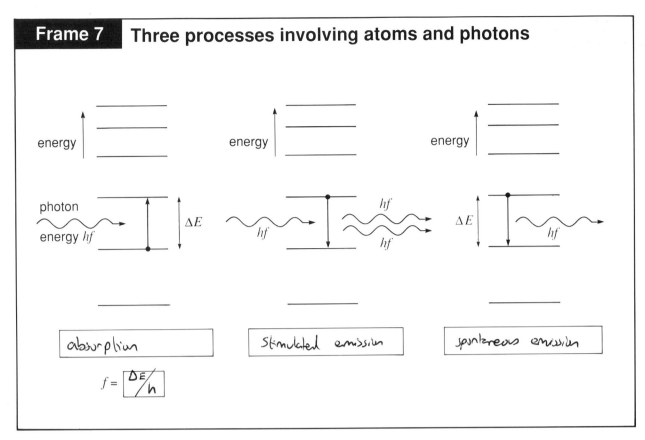

absorption — stimulated emission — spontaneous emission

$$f = \Delta E / h$$

Frame 8: Absorption versus stimulated emission

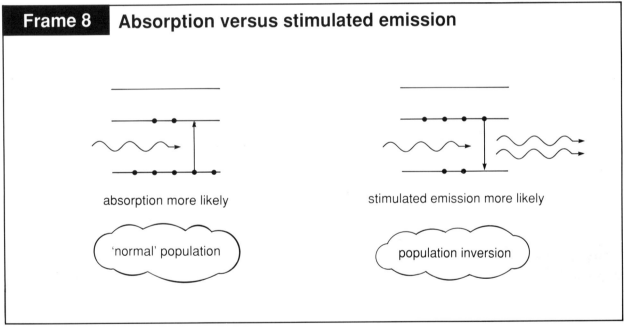

absorption more likely — 'normal' population

stimulated emission more likely — population inversion

Frame 9: Stimulated emission in the helium–neon laser

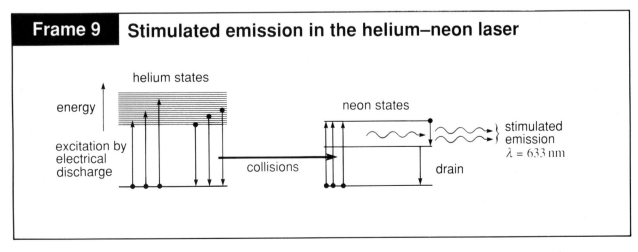

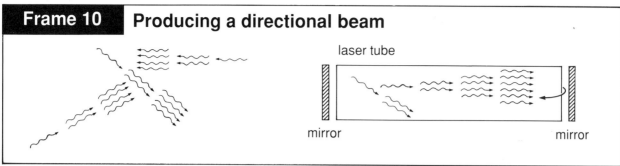

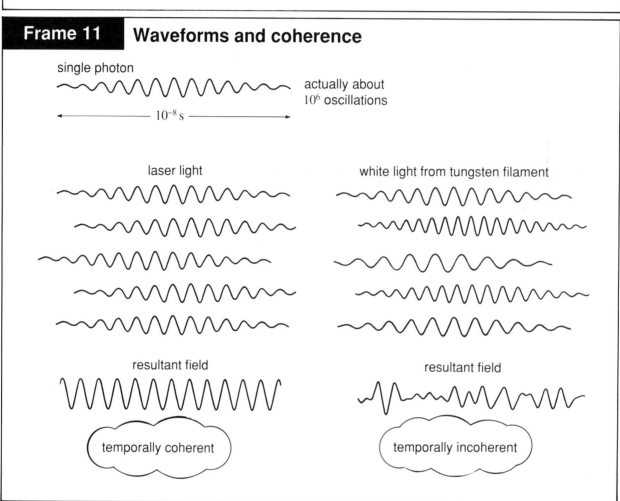

3.1 Summary of Section 3

1 The three major processes in which photons interact with matter are absorption (Frame 4), spontaneous emission (Frame 2) and stimulated emission (Frame 6).

2 The photon energy hf is equal to the difference ΔE between the energy of the atom before and after the absorption/emission process:

$$hf = \Delta E.$$

3 Light from a tungsten filament and from the top of the laser tube is produced by spontaneous emission. The continuous range of energies in the solid filament give rise to a continuous spectrum; the widely separated energies of states in a gas give rise to the line spectra from the laser tube.

4 The laser beam is produced by stimulated emission between two levels for which a population inversion has been established.

5 Mirrors at the ends of the laser tube enhance stimulated emission along the axis of the tube, and a fraction of the light passes through one of the mirrors to produce the highly directional laser beam.

6 The waveform of a laser beam has a regular, sinusoidal profile, and therefore is temporally coherent. White light from a filament has a very irregular waveform, which means that it is temporally incoherent.

SAQ 4 (a) Name the three major processes involving photons that take place in solids, liquids or gases.

(b) Which of these processes is mainly responsible for the light emitted by (i) a laser, and (ii) a tungsten lamp? Which (if any) of the processes named in (a) do *not* occur in each of these two light sources?

SAQ 5 Explain what is meant by population inversion. Why is it an essential prerequisite for laser action?

4 The addition of waves

In Units 1 & 2, you saw that waves add, or superpose, in a very simple way: when two or more waves are travelling through a point in space, the resultant displacement at that point is the sum of the displacements that each individual wave would produce on its own. This is known as the **principle of superposition**, and I used this principle at the end of the audiovisual sequence when discussing the superposition of light from a large number of atomic transitions to produce the resultant light field from a small region of a filament. In other situations, we may be interested in adding light waves from different sources, light waves with the same or different frequencies, or light waves with a broad frequency spectrum. In all of these cases, though, the same principle of superposition applies: the total field is the sum of the individual light fields.

Another important point made in Units 1 & 2 was that our eyes, photographic film and other detectors are unable to respond to the very high frequency oscillations (above 10^{14} Hz) of the field E associated with light waves. Instead they respond to the light **intensity** I (energy flux, watts per square metre), which is proportional to the time average of the square of the field, that is

$$I \propto \langle E^2 \rangle. \tag{6}$$

Since it is the intensity of the light that is important at the detection stage, it is important to understand how the intensity at a point is related to the intensities of the different waves that are travelling through that point. You might be tempted to think that the total intensity must be the sum of the individual intensities, but this is *not* always the case. The numerical example in the following ITQ should alert you to the need for caution.

ITQ 10 Suppose $y = x^2$, and two values of x are $x_1 = 3$ and $x_2 = 1$.
(a) What are the corresponding values of y_1 and y_2?
(b) Suppose $x_r = x_1 + x_2$, and $y_r = x_r^2$. Is y_r equal to $y_1 + y_2$?

In spite of the warning from this ITQ, it turns out that in the majority of cases, we *can* simply add individual intensities to get the resultant intensity. It is only in special circumstances that this simple addition is not valid, and in this section I will make it clear what these special circumstances are. First, though, we will look at the addition of two waves quite generally.

4.1 The general case

Suppose that two waves pass through a certain point in space. How is the resultant intensity due to these waves related to the intensities of the individual waves?

I will represent a wave by the Greek letter ψ (pronounced psi), which depends on both position and time. For the sake of simplifying the algebra that follows, I will assume that ψ is proportional to the field E of the wave, and that the constant of proportionality is chosen in such a way that the intensity I of the wave is related to ψ by

$$I = \langle \psi^2 \rangle. \tag{7}$$

Now according to the principle of superposition, the resultant wave ψ_r at a point due to two waves, ψ_1 and ψ_2, is

$$\psi_r = \psi_1 + \psi_2. \tag{8}$$

At any instant of time, the resultant wave ψ_r at the point in question is simply the sum of the individual waves, ψ_1 and ψ_2, at that instant at that point. This superposition is illustrated in Figure 17(a)–(c). The intensity corresponding to each of the waves ψ_1, ψ_2 and ψ_r can be determined graphically using the definition in equation 7, and this has been done in Figure 17(d)–(f). In each

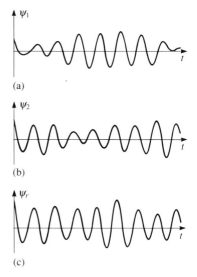
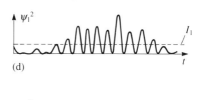
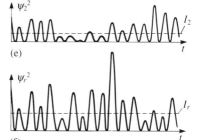

Figure 17 Superposition of two waves, (a) ψ_1 and (b) ψ_2, leads to (c) a resultant wave ψ_r, where at each instant of time $\psi_r = \psi_1 + \psi_2$. (d)–(f) show ψ_1^2, ψ_2^2 and ψ_r^2, respectively; the average values in each case are indicated by the broken lines and correspond to the intensities I_1, I_2 and I_r, respectively.

case, the horizontal broken line represents the average value, over time, of the square of the wave and this corresponds to the intensity. Thus

$$I_1 = \langle \psi_1^2 \rangle; \quad I_2 = \langle \psi_2^2 \rangle; \quad I_r = \langle \psi_r^2 \rangle. \tag{9}$$

To see how I_r is related to I_1 and I_2, we first note that

$$I_r = \langle \psi_r^2 \rangle = \langle (\psi_1 + \psi_2)^2 \rangle.$$

Expanding the squared term,

$$I_r = \langle \psi_1^2 + \psi_2^2 + 2\psi_1\psi_2 \rangle.$$

Now the time-average of the sum of a number of terms is always equal to the sum of the time-averages of the individual terms, that is

$$\langle A + B + C \rangle = \langle A \rangle + \langle B \rangle + \langle C \rangle.$$

Thus

$$I_r = \langle \psi_1^2 \rangle + \langle \psi_2^2 \rangle + \langle 2\psi_1\psi_2 \rangle,$$

and using the relationships in equation 9, this simplifies to

$$I_r = I_1 + I_2 + 2\langle \psi_1\psi_2 \rangle. \tag{10}$$

It is important to note that this result is *always* true. No assumptions were made about the forms of the individual waves, ψ_1 and ψ_2, in the derivation of equation 10. The waves could be sinusoidal, with the same temporal frequency or different frequencies, or the waves could have any arbitrary profiles. Whatever the forms of the two waves, the resultant intensity I_r is always the sum of the intensities I_1 and I_2 of the individual waves, *plus* the term $2\langle \psi_1\psi_2 \rangle$. This latter term involves the **cross-term** $\psi_1\psi_2$, so called because it is the product of the two individual wave profiles, and it is this term that leads to differences between the ways that intensities combine.

Of course, often we are interested in the addition of more than two waves, and fortunately the expression in equation 10 is easily extended to cover any number of waves. Suppose that we want to add waves $\psi_1, \psi_2, \psi_3, \cdots$.

Then

$$\psi_r = \psi_1 + \psi_2 + \psi_3 + \cdots \tag{11}$$

and

$$\begin{aligned} I_r &= \langle (\psi_1 + \psi_2 + \psi_3 + \cdots)^2 \rangle \\ &= \langle \psi_1^2 + \psi_2^2 + \psi_3^2 + \cdots + 2\psi_1\psi_2 + 2\psi_1\psi_3 + 2\psi_2\psi_3 + \cdots \rangle \\ &= I_1 + I_2 + I_3 + \cdots + 2(\langle \psi_1\psi_2 \rangle + \langle \psi_1\psi_3 \rangle + \langle \psi_2\psi_3 \rangle + \cdots). \end{aligned} \tag{12}$$

Thus the resultant intensity is the sum of all of the individual intensities, plus cross-terms involving each pair of waves.

Again, it is the role of the cross-terms that is crucial, and in the next few sections we will look in turn at the behaviour of the cross-terms for a variety of pairs of light waves.

4.2 Adding waves with the same temporal frequency

We will first consider the addition of two waves that have the same temporal frequency (and therefore the same wavelength, and the same colour). This is a case that you met in Units 1 & 2, and possibly you may also have met it in other physics courses. It is the case that leads to constructive and destructive interference.

As with the addition of any two waves, the resultant intensity I_r is given by the general expression that I derived in the previous section:

$$I_r = I_1 + I_2 + 2\langle \psi_1 \psi_2 \rangle. \qquad \text{(Eq. 10)}$$

When the two waves ψ_1 and ψ_2 have the same temporal frequency, the average value of the cross-term $\psi_1 \psi_2$ depends critically on the phase difference $\Delta\phi$ between the two waves, and this is shown in Figure 18 for the special case in which the amplitudes of the waves are equal. When the two waves are in phase ($\Delta\phi = 0$), the cross-term is always positive (Figure 18(a)). Conversely, when the waves are exactly out of phase ($\Delta\phi = \pi$), the cross-term is always negative (Figure 18(e)), since when one wave is positive the other is negative and vice versa. Midway between these two extremes is the case where the phase difference is $\pi/2$ (Figure 18(c)), and here the cross-term is as often positive as it is negative.

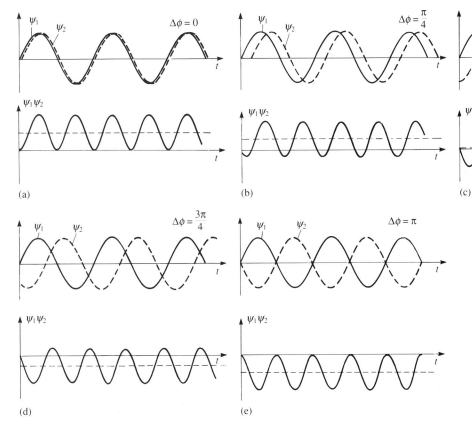

Figure 18 Two waves ψ_1 and ψ_2 with the same temporal frequency and the same amplitude, but with various phase differences. (a) $\Delta\phi = 0$, waves exactly in step; (b) $\Delta\phi = \pi/4$; (c) $\Delta\phi = \pi/2$; (d) $\Delta\phi = 3\pi/4$; (e) $\Delta\phi = \pi$, waves exactly out of step. The cross-terms $\psi_1 \psi_2$ are plotted at the bottom of each part of the figure, and the average values are indicated by broken lines. Note that at any particular time, the value plotted for the cross-term is equal to the product of the values plotted for ψ_1 and ψ_2 at that time.

ITQ 11 What is the intensity I_r that results from superposition of the two waves ψ_1 and ψ_2 in Figure 18(a)? Express the answer in terms of the intensity I_1 of the individual waves. (Bear in mind that $\psi_1 = \psi_2$.)

Superposition of the waves in Figure 18(a) corresponds to **constructive interference**: note that here the resultant intensity is twice as large as the sum of the individual intensities.

In the case where $\Delta\phi = \pi$ (Figure 18(e)), $\psi_1 = -\psi_2$, and so

$$\langle \psi_1 \psi_2 \rangle = -\langle \psi_1^2 \rangle = -I_1.$$

Thus

$$I_r = I_1 + I_2 + 2\langle \psi_1 \psi_2 \rangle = I_1 + I_1 - 2I_1 = 0.$$

Here the two waves cancel completely and the intensity is zero: we have complete **destructive interference**.

It is worth pointing out that the resultant intensities just deduced from equation 10 for the special cases of complete constructive interference ($\Delta\phi = 0$)

and complete destructive interference ($\Delta\phi = \pi$) can equally well be deduced by adding the two waves, and then working out the intensity of the resultant wave. Thus:

for constructive interference: $\quad \psi_r = \psi_1 + \psi_2 = \psi_1 + \psi_1 = 2\psi_1$

so $\quad I_r = \langle \psi_r^2 \rangle = \langle (2\psi_1)^2 \rangle = 4\langle \psi_1^2 \rangle = 4I_1.$

for destructive interference: $\quad \psi_r = \psi_1 + \psi_2 = \psi_1 - \psi_1 = 0$

so $\quad I_r = \langle \psi_r^2 \rangle = 0.$

For other phase differences besides $\Delta\phi = 0$ and $\Delta\phi = \pi$, the resultant intensity lies somewhere between the extreme values of $4I_1$ and zero. The special cases of $\Delta\phi = \pi/2$ and $\Delta\phi = 3\pi/2$ result in cross-terms with average values that are equal to zero (Figure 18(c)), so in these cases equation 10 leads to

$$I_r = I_1 + I_1 + 0 = 2I_1.$$

The other examples in Figure 18(b), (d) show how partial constructive interference occurs when the cross-term is more positive than negative, and how partial destructive interference occurs when the cross-term is more negative than positive.

The important point to remember here is that when adding waves with the same temporal frequency, the resultant intensity is *not* the sum of the individual intensities (except in special cases when the phase difference between the waves is $\pi/2$ or $3\pi/2$). The time-averaged value of the cross-term varies from $-I_1$ when $\Delta\phi = \pi$ to $+I_1$ when $\Delta\phi = 0$, and it is the variation of this cross-term with phase difference that is at the root of all interference phenomena.

4.3 Adding waves with different temporal frequencies

At first sight, the addition of waves with different temporal frequencies might seem more complicated than the addition of waves with the same frequency. After all, the temporal frequency difference means that the waves will be in step at some instants of time and out of step at other times. However, the fact that the waves are sometimes in step and sometimes out of step leads to a simplification of the expression for the resultant intensity.

To see how this simplification comes about, we again start with the basic relationships for superposition of two waves:

$$\psi_r = \psi_1 + \psi_2 \qquad \text{(Eq. 8)}$$

and

$$I_r = I_1 + I_2 + 2\langle \psi_1 \psi_2 \rangle. \qquad \text{(Eq. 10)}$$

Remember, these two equations are always true, whatever the nature, or whatever the profile, of the waves that we are considering. Once again it is the behaviour of the cross-term $\psi_1 \psi_2$ that is the key to finding the resultant intensity.

Figure 19 shows a specific example of two waves ψ_1 and ψ_2, with different frequencies, and the corresponding cross-term $\psi_1 \psi_2$ is plotted at the bottom of the figure. Note that the maximum positive values of the cross-term occur where the peaks of the two waves coincide, or where the troughs of the two waves coincide. Conversely, the maximum negative values occur where a peak of one wave coincides with a trough of the other. Between these two extremes, the cross-term oscillates from positive to negative. However, the exact nature of the variation of the cross-term is unimportant. What is important is that the average value over time, $\langle \psi_1 \psi_2 \rangle$, is zero, because the positive values are exactly balanced by the negative values.

So since the average value of the cross-term is zero for waves with different frequencies, we can simplify equation 10 by omitting the last term:

two waves with different frequencies: $\quad I_r = I_1 + I_2.$ \qquad (13)

☐ Would you expect to observe interference effects when waves with different frequencies are superposed?

■ No interference effects would be apparent: the cross-term $\psi_1 \psi_2$ is responsible for interference effects, and this term is absent from equation 13 because its time-average value is zero.

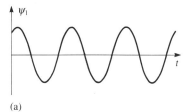

(a)

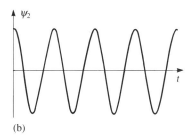

(b)

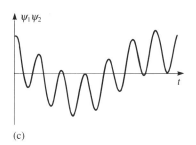

(c)

Figure 19 Two waves, ψ_1 and ψ_2, with different frequencies. The cross-term $\psi_1 \psi_2$ has its maximum positive values where the two waves are in step and its maximum negative values where the waves are exactly out of step.

Thus when the two waves being added have different frequencies, the resultant intensity is simply the sum of the intensities of the individual waves.

The expression in equation 13 is valid whatever the values of the two frequencies, as long as they are different. The rate at which the two waves change from being in step to being out of step is equal to the difference between the frequencies of the waves. The smaller the difference between the two frequencies, the longer it will take for the two waves to change from in step to out of step. However, even if the difference between the frequencies of two light waves were only one part in a billion (10^9), this difference would amount to $10^{-9} \times (5 \times 10^{14}\,\text{Hz})$, say, which is $5 \times 10^5\,\text{Hz}$. This means that the waves would change from being in step to out of step 5×10^5 times per second, which is far faster than the eye or film can respond. Thus even in this case, interference effects would not be visible, and the resultant intensity would be the sum of the individual intensities.

To conclude this section, I should point out that equation 13 can be extended to deal with any number of waves with different frequencies. In the general expression in equation 12, all of the cross-terms will be zero if ψ_1, ψ_2, ψ_3, etc. are waves with different frequencies. Thus

$$\text{many waves with different frequencies:} \quad I_r = I_1 + I_2 + I_3 + \cdots. \tag{14}$$

4.4 Adding waves from different source points

So far, we have been considering superposition of monochromatic waves, with either the same frequency or different frequencies. However, in real life we are often concerned with superposition of more complex waveforms, such as the light emitted from different regions of a tungsten filament (Figure 20). We will now consider what the resultant intensity is in cases like this.

Let me emphasize once again that the general equations for obtaining the resultant waveform and the resultant intensity apply to this situation, as they do to all others. Thus

$$\psi_r = \psi_1 + \psi_2 \tag{Eq. 8}$$

and

$$I_r = I_1 + I_2 + 2\langle \psi_1 \psi_2 \rangle. \tag{Eq. 10}$$

However, we will look at the average value of the cross-term again, to see if equation 10 can be simplified.

Now as I pointed out when discussing Frame 11 of the audiovisual sequence, the light emitted from a tungsten filament is the superposition of light arising from many atomic transitions in the metal. These transitions corresponds to a range of different temporal frequencies, and this frequency range is determined by the temperature of the filament. What is more, the transitions occur at random times, and consequently the amplitude and phase of the resultant light wave from a small region of the filament change in an unpredictable way.

Thus although electronic transitions leading to light emission occur on average at the same rate in the similar-sized regions 1 and 2 of the filament shown in Figure 20, and although the range of temporal frequencies of the light from the two regions is the same, nevertheless the random timing of the emission processes means that the temporal variations of the waves from the two regions are quite different. This difference is illustrated schematically in Figure 21. The two waves shown in (a) and (b) are superficially quite similar — the average periods are similar and the amplitudes are similar. But though the waves are approximately in step in the regions indicated by arrows, the phase changes in each wave mean that only a short time elapses before they are out of step. The effect that this has on the cross-term is shown in Figure 21(c). The variations of the phase difference between the two waves means that the cross-term is negative as often as it is positive, and so its time-averaged value is zero. Thus we can again simplify equation 10 for the resultant intensity:

$$\text{waves from different sources:} \quad I_r = I_1 + I_2. \tag{15}$$

This is exactly the same result as we obtained in the previous section for the resultant intensity when waves with different frequencies are superposed, and again it is valid because $\langle \psi_1 \psi_2 \rangle$ is zero. The absence of the cross-term from equation 15 means that no effects due to constructive and destructive interference occur when waves from different source points are superposed.

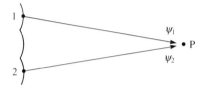

Figure 20 What is the resultant intensity I_r at point P due to waves ψ_1 and ψ_2 that originate from regions 1 and 2 on the filament?

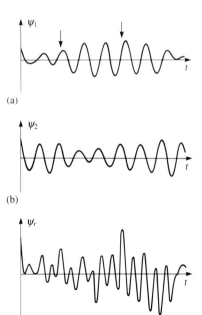

Figure 21 (a), (b) Schematic representation of light waves ψ_1 and ψ_2 from two points on a tungsten filament. The waves are approximately in step in the regions indicated by arrows. (c) The cross-term $\psi_1 \psi_2$; over a sufficiently long time, the average value of $\psi_1 \psi_2$ will be zero.

Now the average value $\langle \psi_1 \psi_2 \rangle$ of the cross-term is essentially a measure of the **correlation** between the two waves. When the two waves ψ_1 and ψ_2 have different frequencies, or when they arise from different source points, there is no correlation between them. Positive values of ψ_1 occur with positive values of ψ_2 just as often as with negative values of ψ_2, and so the average value of $\psi_1 \psi_2$ is zero. On the other hand, when two waves have the *same* frequency, there is a correlation between them, and in this case $\langle \psi_1 \psi_2 \rangle$ has a value that depends on the phase difference between the waves.

But correlations between two waves, together with the consequent interference effects, are also possible in some situations where the waves are not monochromatic. It is, in fact, possible to generate two source points that are correlated. This can be done by using a single point source to produce two (or more) spatially separated secondary sources. Since these secondary sources have a common origin, the light waves that they produce will be correlated.

Three examples of ways that such correlated sources can be produced are shown in Figure 22. In the first of these, Figure 22(a), the wavefront from a point source S_1 (not necessarily monochromatic) is divided by a semi-reflecting mirror, known as a **beam-splitter**. The beam-splitter has only a thin layer of reflecting coating on its surface, so part of the incident wave is transmitted through it and part is reflected. The reflected part is redirected by a second mirror towards region P, where it overlaps with the wave that was transmitted by the beam-splitter. From P, it appears that waves are arriving from sources S_1 *and* S_2. However, the wave from the virtual source S_2 is actually emitted from the real source S_1, so the wave arriving at P from S_2 is often highly correlated with the wave from S_1.

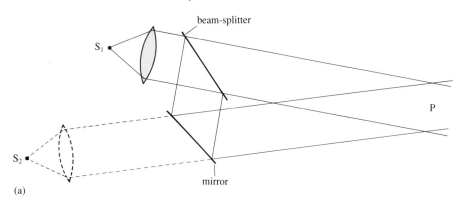

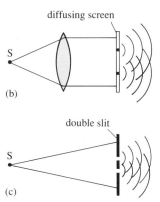

Example (b) in Figure 22 uses a point source S and a lens to illuminate a ground-glass screen. Points on the screen will act as secondary sources of illumination, and again the light waves from these secondary sources will be correlated because they originate from the same primary source S. The third example in Figure 22 is the well-known double slit. The two slits act as secondary sources, but the light from each of them arises from a single point source. This again means that the illumination from the two slits is correlated.

Figure 22 Three ways to produce correlated sources. (a) Use a beam-splitter; light from the virtual source S_2 is correlated with light from the real source S_1. (b) Use a ground-glass screen. (c) Use a double slit illuminated with a point source.

In all three of these examples, the average value, $\langle \psi_1 \psi_2 \rangle$, of the cross-term may not be zero because of correlation between the two waves. When calculating the intensity at a point, it is therefore necessary to use the full expression for the resultant intensity:

$$I_r = I_1 + I_2 + 2\langle \psi_1 \psi_2 \rangle. \quad \text{(Eq. 10)}$$

Consequently, interference effects may be observed, just as they are when monochromatic waves with the same frequency are superposed.

4.5 Summary of Section 4

1 According to the *principle of superposition*, when two waves are travelling through a point in space, the resultant displacement (or field) at that point is the sum of the displacements (or fields) that each individual wave would produce on its own;

$$\psi_r = \psi_1 + \psi_2. \quad \text{(Eq. 8)}$$

2 Light detectors respond to *intensity I* (energy flux, watts per square metre);

$$I = \langle \psi^2 \rangle. \quad \text{(Eq. 7)}$$

3   The resultant intensity due to two waves is the sum of the intensities of the individual waves plus twice the time-average of the cross-term:

$$I_r = I_1 + I_2 + 2\langle \psi_1 \psi_2 \rangle. \qquad \text{(Eq. 10)}$$

4   If the waves being added have the same temporal frequency (same wavelength), then the waves are correlated and the time-average of the cross-term is generally not zero; interference effects can therefore be observed.

5   Waves that have different frequencies, or waves emitted from different source points, are uncorrelated, so the time-averages of any cross-terms are zero. In these cases,

$$I_r = I_1 + I_2, \qquad \text{(Eq. 13, 15)}$$

and interference effects are not observed.

6   It is possible to produce two (or more) waves from a single point source; in such cases, the waves may be correlated, so the full expression for the resultant intensity must be used (equation 10 above), and interference effects may be observed.

SAQ 6   Two sinusoidal waves, ψ_1 and ψ_2, with identical wavelengths, pass through a particular region of space. In this region, the amplitude of the field of the first wave is 1.5 times greater than the amplitude of the second. Calculate the resultant intensity due to the two waves at points in this region where the phase difference $\Delta\phi$ between the waves is (a) 0, (b) π, and (c) $\pi/2$. Express your answers in terms of the intensity I_2 of the second wave.

SAQ 7   In which of the following situations could constructive and destructive interference be observed?

(a)   A region of space illuminated by a red laser beam and a green laser beam.

(b)   A region illuminated by two independent point sources, each covered with a red filter.

(c)   A region behind a diffraction grating, which is illuminated by an expanded laser beam.

5   Temporal coherence

5.1   Introducing coherence

Your experiments earlier in this unit demonstrated that the nature of a diffraction pattern depends on the illumination reaching the object, as well as depending on the object itself. For example, when you altered the illumination by changing the size of the source, or by changing the range of wavelengths present, the diffraction pattern changed quite dramatically. This dependence of the diffraction pattern on the illumination has important applications, for it allows us to deduce information about the illumination from observations of the diffraction pattern.

In the second half of this unit, I want to concentrate on a number of ways of characterizing the illumination *at the object*. The important concept that I will introduce to characterize the illumination is known as *coherence*. This term has a precisely defined technical meaning, but this is closely related to the meaning of the word in everyday conversation. If we say that an argument, a speech or a policy is coherent, then we mean that the whole thing fits together, it's all connected, the various parts are consistent. In the same way, illumination is coherent if the wave at one point is consistent with the wave at another point, and as you will see, the consistency requires a correlation between the phases of the waves at the two points.

As you might guess, the coherence of illumination is dependent both on spectral bandwidth and on range of inclinations, but it is often possible to separate these two effects, just as you did with the Home Kit experiments. The spectral bandwidth determines what is known as the *temporal coherence* of the illumination and the range of inclinations determines the *spatial coherence*. These two types of coherence will be discussed in this section and in Section 6, respectively. You will see that measurements of the coherence properties can provide important information, ranging from the size of a star to the composition of the exhaust gases from a car engine.

5.2 Temporal coherence and bandwidth

Temporal coherence is related to the spectral bandwidth of the illumination, that is to the range of temporal frequencies (or wavelengths) that are present in the illumination It is *not* related to the size of the source or to the distance between source and object. So, to simplify discussion of temporal coherence in this section, we will deal only with very distant point-like sources of illumination; this means that effects due to the size of the source can be neglected, and the illumination can be regarded as plane waves travelling in a single direction.

☐ How would you determine the range of temporal frequencies in the illumination from a point source?

■ You have essentially done this already in the Home Kit experiments in Section 2.5. When unfiltered light from the tungsten filament illuminated a grating, each spatial frequency component produced a rainbow of colour in the diffraction pattern, ranging from violet and blue nearer the optical axis out to orange and red. By eye, you could see that there was a continuous distribution of temporal frequencies throughout the visible range, and an estimate of the range of frequencies could be made by measuring the distances from the optical axis of the blue and red ends of a spectrum. It would also be possible to use a small photometer to measure how the intensity varies across the spectrum in the diffraction pattern. This is the principle behind many spectrophotometers that are used in industrial and research laboratories.

When illumination has a continuous spectrum, such as that shown in Figure 23, the spectrum can be rather crudely characterized by two parameters. One pair that we can use is the average wavelength λ_{av} and the range of wavelengths $\Delta\lambda$. Alternatively, we could use the average frequency f_{av} and the range of frequencies Δf. The range of wavelengths $\Delta\lambda$ (or frequencies Δf) is referred to as the wavelength (or frequency) **bandwidth**, and it is conventional to measure it between points where the intensity is 50% of its maximum value, as indicated in Figure 23.

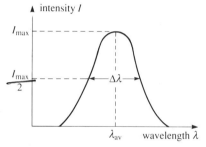

Figure 23 A continuous spectrum. The average wavelength is λ_{av}, and the bandwidth $\Delta\lambda$. It is conventional to measure the bandwidth between the points at which the intensity drops to half of the maximum value.

The relationship between these two pairs of parameters depends on the shape of the spectrum and on the relative magnitudes of $\Delta\lambda$ and λ_{av}. However, for our purposes it is sufficiently accurate to use the following approximate relations which are valid when $\Delta\lambda \lesssim \lambda/2$:

$$f_{av} \approx c/\lambda_{av} \qquad (16)$$

and

$$\frac{\Delta f}{f_{av}} \approx \frac{\Delta\lambda}{\lambda_{av}}. \qquad (17)$$

It is the bandwidth $\Delta\lambda$ (or Δf) that determines the temporal coherence of the illumination. Laser light has a very *narrow* bandwidth, and it therefore has very high temporal coherence, as I pointed out when discussing Frame 11 of the AV sequence. As the bandwidth is *increased*, the temporal coherence is *reduced*. Thus a narrow bandwidth implies high temporal coherence, whereas a wide bandwidth implies low temporal coherence, and this relationship is illustrated in Figure 24.

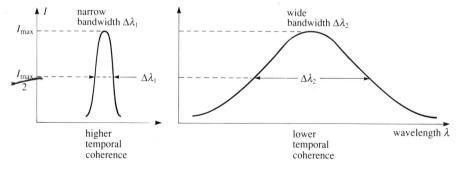

Figure 24 There is a reciprocal relationship between bandwidth and temporal coherence.

The reciprocal relationship between bandwidth and temporal coherence is an important one to remember. However, it doesn't give any insight into why the term coherence is used to describe this property of illumination. I can throw some light on this, metaphorically speaking, by considering another method of measuring temporal coherence.

5.3 Measuring temporal coherence with a double slit

Suppose that a double slit mask was illuminated with a plane-wave beam, as shown in Figure 25. The experiments in Section 2.5 should have convinced you that the diffraction pattern in the back focal plane on the lens would depend on the bandwidth of the illumination. But the diffraction pattern also provides a *quantitative* measure of the bandwidth, and therefore of the temporal coherence, as you will see in the next two sections.

You may well have come across the double slit in previous physics courses, since it is frequently used to teach concepts of superposition, or interference, of light. In that context, the diffraction pattern would most probably have been discussed using the Huygens approach. But you also met the double slit when studying Units 3 & 4, where it was discussed using the Fourier approach. In fact, these two approaches provide complementary views of temporal coherence. I shall therefore make use of both of them, starting with the Fourier approach.

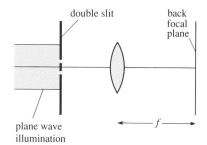

Figure 25 The lens forms a diffraction pattern in its back focal plane. The temporal coherence of the illumination can be determined from measurements made on the diffraction pattern.

5.3.1 The Fourier approach to temporal coherence

The transform pairs in Figure 26 are Fourier mates that you met in the previous units. The pair at the top show that the Fourier transform of a cosinusoidal amplitude transmittance is three delta functions, and the bottom pair show that the Fourier transform of two delta functions is a cosinusoidal function. Note that these pairs are not quite the reverse of each other: in the first pair, the cosinusoidal distribution has a constant amplitude added to it so that it is always positive, and this constant amplitude produces the central (zero-frequency) delta function. The constant amplitude component and the associated delta function are not present in the bottom pair.

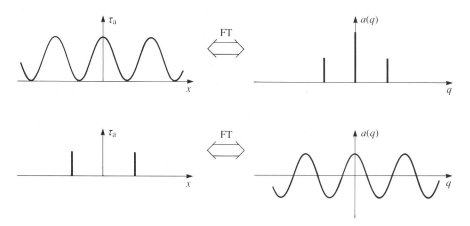

Figure 26 Two amplitude transmittance profiles (left) and their Fourier transforms (right).

It is the bottom pair in Figure 26 that will concern us here, because the two delta functions represent the transmittance of a double slit, and the cosinusoidal distribution represents the diffraction field produced in the back focal plane of a convex lens. Note that this is a cosinusoidal *field*; the intensity is proportional to the square of this graph, so the intensity will also be cosinusoidal, but it will always be positive and it will have *double* the spatial frequency.

> **ITQ 12** What assumption is being made about the slits if their transmittance is represented by the delta functions shown at the bottom left in Figure 26? What would be the effect on the Fourier transform of using a more realistic transmittance?

Now the Fourier pairing shown at the bottom of Figure 26 corresponds to a double slit illuminated with a *monochromatic* plane wave travelling parallel to the optical axis. But our present concern is with illumination that is composed of a *range* of wavelengths. In this situation, each of the wavelengths present — or each temporal frequency, if you prefer — gives rise to a cosinusoidal field distribution in the back focal plane of a lens. However, the distance between the peaks of each of these cosinusoidal fields is proportional to the wavelength of the light, because distances in the diffraction pattern are proportional to wavelength (remember, $s = fq\lambda$ when the diffraction angle is small). This scaling of the fields in the back focal plane is shown in Figure 27.

The resultant effect of these overlapping cosinusoidal fields, each corresponding to a different temporal frequency, depends on the detector that is used. Your eyes can distinguish the different coloured fringes, but black and white film, or a photometer like the one in the Home Kit, responds to the resultant intensity and does not distinguish different colours.

☐ How would you determine graphically the resultant intensity due to the three fields shown in Figure 27?

■ The resultant intensity at each point is simply the *sum of the intensities* due to each of the three different wavelengths at that point. Remember, in Section 4.3, I showed that when superposing light with different wavelengths (different temporal frequencies), the total intensity is the sum of the individual intensities.

Figure 28(a)–(c) shows the three intensity patterns that correspond to the cosinusoidal fields in Figure 27, and the result obtained by adding the three intensity graphs is shown in Figure 28(d). It is clear that the cosinusoidal fringes obtained with monochromatic light get blurred out when several wavelengths are present. This effect is even more dramatic when many wavelengths throughout the visible range are present in the illumination: only a few fringes are then visible before they disappear into a uniform background intensity. It is this disappearance of the fringes away from the centre of the diffraction pattern of a double slit that can be related to the temporal coherence of the illumination, or alternatively to its bandwidth.

The relationship between temporal coherence and fringe pattern is illustrated in Figure 29, which shows double slit diffraction patterns produced with three different types of illumination — laser illumination, illumination with wavelengths between 550 nm and 650 nm, and illumination covering the whole visible range (white light). Highly coherent laser illumination has a very narrow bandwidth and the fringes remain sharply defined away from the centre of the pattern. White light, with its low temporal coherence and broad bandwidth, produces a diffraction pattern with only a few fringes. The illumination with intermediate temporal coherence and intermediate bandwidth produces an intermediate number of fringes. So, illumination with high temporal coherence generates a large number of well-defined fringes in the double slit diffraction pattern, whereas low temporal coherence results in few fringes.*

Now it is possible to show (though I will not do so here) that the number of fringes that are visible is related to the ratio of bandwidth $\Delta\lambda$ to average wavelength λ_{av} in the following way:

$$n \approx \lambda_{av}/\Delta\lambda, \qquad (18)$$

where n is the number of bright fringes between the centre of the diffraction pattern and the position where the fringes blur out to uniform intensity.

ITQ 13 White light from a point source is passed through a green filter before it illuminates a double slit. Eight bright fringes are observed on either side of the central fringe in the diffraction pattern. Assuming that the slits are extremely narrow, and that the average wavelength of the green light transmitted by the filter is 530 nm, what is the bandwidth of the illumination?

5.3.2 The Huygens approach to temporal coherence

We will now look at double slit diffraction using the Huygens approach. The reason for breaking away from the Fourier approach, which has dominated the discussion in Units 3 & 4, is that the Huygens alternative gives additional insight into temporal coherence. Indeed, it will make it clearer why this term is used to describe a characteristic property of illumination.

(a) Monochromatic illumination

To start with, consider what happens when a double slit is illuminated with *monochromatic* plane-wave illumination, travelling parallel to the optical axis. According to Huygens' model, the two slits act as secondary sources,

---

* Note that this effect of temporal coherence is quite separate from the superficially similar effect caused by finite-width slits, which was discussed in ITQ 12.

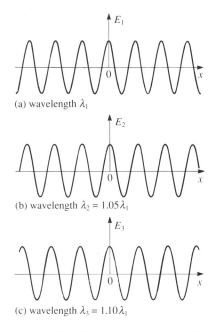

(a) wavelength λ_1

(b) wavelength $\lambda_2 = 1.05\lambda_1$

(c) wavelength $\lambda_3 = 1.10\lambda_1$

Figure 27 Three cosinusoidal field patterns produced in the back focal plane of a lens when light with three different wavelengths is diffracted by a double slit. The wavelengths, and therefore the distances between the peaks of the fields, are in the ratio 1.00 : 1.05 : 1.10.

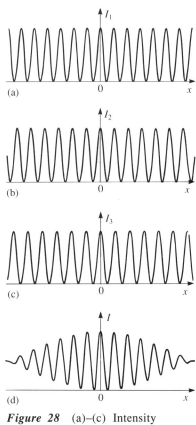

Figure 28 (a)–(c) Intensity patterns corresponding to the fields shown in Figure 27(a)–(c). (d) The resultant intensity due to the three intensity distributions in (a)–(c). At any value of x, the resultant intensity is the sum of the intensities shown in (a), (b) and (c) at that value of x. Note that the scaling of the intensity axis is different in (d) from that in (a)–(c).

with wavefronts spreading out from each slit in all directions, as shown in Figure 30. As usual, the lens forms a diffraction pattern in its back focal plane. The light arriving at a particular point in the diffraction plane, such as point A, will have travelled along parallel paths between the slits and the lens, indicated by broken lines, before being brought together by the lens to point A.

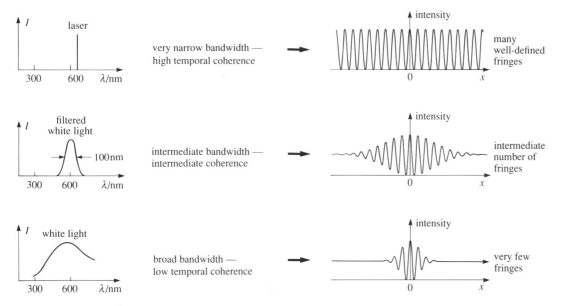

Figure 29 Bandwidth and temporal coherence (left) and the fringe pattern produced by a double slit (right) for three types of illumination.

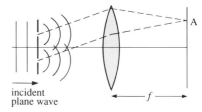

Figure 30 Wavefronts spreading out from a double slit. The broken lines represent two rays, initially parallel, that are brought together at point A in the back focal plane of the lens.

Now the light transmitted by the two slits is in phase, because we are assuming that the initial illumination is a plane wave travelling along the optical axis. But since the two broken line paths shown in Figure 30 are of unequal lengths, the light from the two slits is not necessarily in phase when it reaches point A. We therefore have the possibilities of constructive and destructive interference taking place, as discussed in Section 4.2. According to Huygens' model, it is constructive and destructive interference that accounts for the bright and dark fringes observed in the diffraction pattern. But how do we determine whether light from the two slits is exactly in phase when it arrives at A (i.e. constructive interference, bright fringe), or whether it has the opposite phase (destructive interference, dark fringe), or whether it has some intermediate phase difference?

A direct calculation of the number of wavelengths that correspond to each of the two broken-line paths in Figure 30 would be rather tedious, because it would have to take account of the curved surfaces of the lens and of the different wavelength of the light as it travelled through the glass. Fortunately, we can bypass such a calculation. We simply make use of the fact that a convex lens focuses a plane wave to a point in its back focal plane. This is illustrated in Figure 31(a): the lens converts plane wavefronts to spherical wavefronts that converge to point A in the back focal plane. The important point to note is that there are exactly the same numbers of wavelengths in the path between point S_1 and point A as between point R and point A, *if* S_1 and R are on the same wavefront. If this fact is applied to the double slit problem, as redrawn in Figure 31(b), then again there are the same number of wavelengths along the paths between S_1 and A and between R and A. This means that the path difference between S_1A and S_2A is equal to the small segment of path between S_2 and R. So it is the number of wavelengths between S_2 and R that determines the phase difference between the waves arriving at A from the two slits.

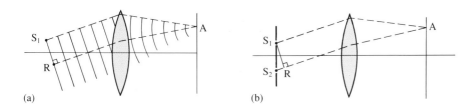

Figure 31 (a) Both S_1 and R are separated from A by the same number of wavelengths. (b) S_2R is the extra distance that light passing through slit S_2 travels to reach A relative to the distance travelled by light passing through S_1.

ITQ 14 Figure 32 shows light waves travelling away from the double slit at four different values of the angle θ.

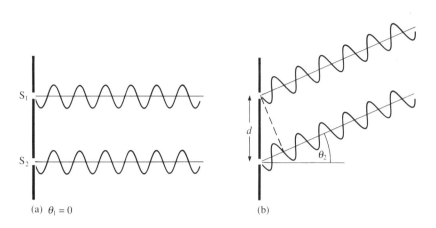
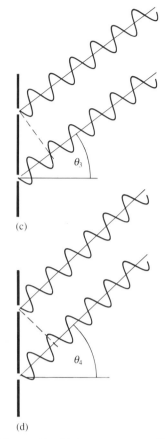

(a) For each of these angles, deduce from the wavetrains illustrated whether the superposition of the light from the two slits in the back focal plane of a lens will lead to complete constructive interference, complete destructive interference or something between the two extremes.

(b) Write down the equation that expresses the condition for complete constructive interference to occur. Your equation should relate the angle θ to the slit spacing d and the wavelength λ.

Now there is an important point implicit in Figures 31 and 32: because light from the two slits has to travel different distances to reach a point in the diffraction plane (except when $\theta = 0°$), one wave will be delayed relative to the other — it will arrive at a later time. For the example shown in Figure 32(b), light from S_2 must travel an extra distance λ to reach the diffraction plane compared with light from S_1. It will therefore be delayed by the time that it takes for light to travel this extra distance, and this delay is given by

$$\text{delay} = \frac{\text{extra distance}}{\text{speed of light}} = \frac{\lambda}{c} = T,$$

where c is the speed of light, and T is the period of the wave. Similarly, the extra distances travelled in the cases shown in Figure 32(c) and (d) are 1.5λ and 1.7λ, respectively, which correspond to time delays of $1.5T$ and $1.7T$, respectively.

The effects of these delays are illustrated in Figure 33, which shows the time dependences of the waves from the two slits. On the left of the figure are the waves at the two slits during a short time interval, and I have marked with stars two peaks of the fields that occur at the same time. On the right of the

Figure 32 Light waves travelling away from a double slit at four different angles. Will these waves interfere constructively or destructively?

Figure 33 Field versus time at the two slits (left), and (a)–(d) at four points in the diffraction pattern corresponding to the angles in Figure 32 (a)–(d). The peaks marked with stars occur simultaneously at the slits, but occur at different times in the diffraction plane.

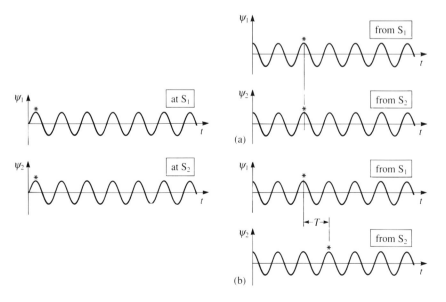
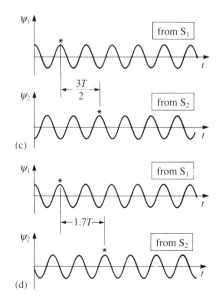

figure, in parts (a)–(d), I have shown the fields at four points in the diffraction plane around the time when the starred peaks arrive; these four parts correspond to the light that initially travelled away from the slits at the four angles illustrated in Figure 32(a)–(d). Thus in Figure 33(a), which corresponds to the $\theta = 0°$ case in Figure 32(a), the two starred peaks in the field occur simultaneously. Because the waves from the two slits travel the same distance in this case, there is no delay of one wave relative to the other, and the two waves are exactly in phase in the diffraction plane. Clearly these waves will interfere constructively, producing a bright fringe at the centre of the diffraction pattern. The second example, in Figure 33(b), shows that the wave from S_2 is delayed by one period (T) relative to the wave from S_1; since the wave from S_2 has to travel an extra distance λ, the starred peak of this wave occurs at a later time than the starred peak from S_1. However, since the delay is exactly one period T, the waves are exactly in phase, and so constructive interference occurs at this point too, producing another bright fringe. In Figure 33(c), the delay is $1.5T$, because the wave from S_2 has to travel a distance 1.5λ further than the wave from S_1. This delay means that the waves at this point in the diffraction plane are exactly out of phase, and so complete destructive interference occurs, which results in a dark fringe. Finally, in Figure 33(d), the delay is $1.7T$, and the interference in this situation will lead to an intensity that is between those for the bright and dark fringes.

So with the monochromatic illumination that we are considering now, the result is clear cut. If the extra distance travelled by light from one slit is an integral number of wavelengths $n\lambda$, so that the extra transit time is $n\lambda/c = nT$, then the two waves are in phase and constructive interference occurs: a bright fringe results. Conversely, if the extra distance is $(n + \frac{1}{2})\lambda$, and the extra transit time $(n + \frac{1}{2})T$, then the two waves will be exactly out of phase in the back focal plane, so that complete destructive interference occurs (i.e. a dark fringe). So you can think of the fringes in the back focal plane as a measure of how well the wave matches up with a delayed version of itself. The continuing sequence of bright and dark fringes obtained with monochromatic light is a result of the sinusoidal waveform: this will always superpose constructively with a version of the same waveform that is delayed by a time interval of nT.

So here is the reason that the term 'temporally coherent' is used to describe monochromatic light. The phase of the waveform remains predictable — or coherent — over long temporal (i.e. time) intervals. If we split a temporally coherent beam of light, and delay one part by making it travel a larger distance before the beams are recombined, then we observe constructive or destructive interference.

(b) Broad bandwidth illumination

Having considered the case of monochromatic illumination, I now want to use the same approach to discuss illumination that contains a range of wavelengths. A typical diffraction pattern obtained with broad bandwidth illumination is shown in Figure 34(a). There are only a few bright and dark fringes on either side of the centre of the pattern, compared with the large number of fringes obtained with monochromatic light. To explain why this is so using the Huygens approach, we have to investigate how the waves from the two slits superpose at various points in the diffraction pattern.

The important points to bear in mind are that (a) the waves leaving the two slits are identical, (b) the waves arriving at the back focal plane have a time shift relative to each other, and (c) this time shift is zero on the optical axis of the double slit but increases with distance from the axis.

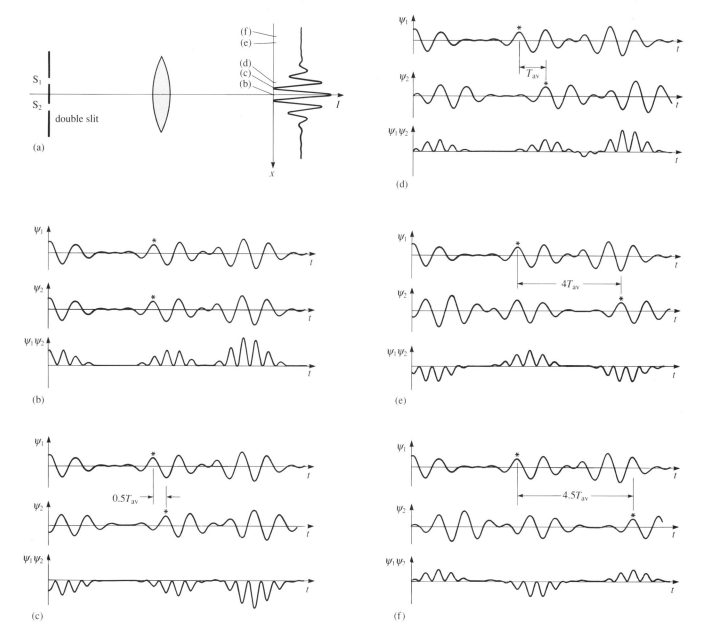

Part (b) in Figure 34 shows the waves ψ_1 and ψ_2 from the two slits at the position of the central bright fringe in the diffraction plane ($\theta = 0°$). Note that the two waves are identical, since we are assuming that both slits are illuminated by a plane wave travelling parallel to the optical axis, and the waves travel the same distance from the two slits to the central point of the diffraction plane. Of course, the waves are *not* sinusoidal because the broad bandwidth means that there are variations in amplitude and phase. Since the two waves are identical, the product $\psi_1\psi_2$ is always positive (as shown at the bottom of Figure 34(b)) and $\langle \psi_1\psi_2 \rangle = \langle \psi_1^2 \rangle = I_1$. So the resultant intensity at this central point is just

$$I_r = I_1 + I_2 + 2\langle \psi_1\psi_2 \rangle \qquad \text{(Eq. 10)}$$
$$= I_1 + I_1 + 2I_1 = 4I_1.$$

Now look at part (d) of Figure 34 (we will discuss part (c) later); here you can see the waves that produce the first bright fringe to one side of the central fringe. The wave ψ_2 has been delayed by time T_{av} (the average period of the light), because of the extra distance λ_{av} that the light from slit S_2 travels to this point relative to the distance travelled by light from S_1. Note that ψ_2 would be identical to ψ_1 if it were shifted to the left of the diagram by an amount T_{av}, that is, if it were shifted earlier in time. Now, because these waves are not perfectly periodic, the delay of ψ_2 by the average period T_{av} means that the two waves are no longer exactly matched. At most times, they are approximately in step, but at some instants they have the opposite sign. This is indicated by the fact that the product $\psi_1\psi_2$ is negative for part of the

Figure 34 (a) Double slit diffraction pattern produced by illumination with a broad bandwidth. (b)–(f) show computer simulated waves, ψ_1 and ψ_2, from the two slits, and the cross-term $\psi_1\psi_2$, at the five points in the diffraction pattern that are labelled (b)–(f). The stars indicate peaks of the waves that occurred at the same instant at the two slits.

time, and it means that $\langle \psi_1\psi_2 \rangle$ is less than I_1. In the illustration shown in Figure 34(d), $\langle \psi_1\psi_2 \rangle \approx 0.6I_1$, and so the resultant intensity is

$$I_r = I_1 + I_2 + 2\langle \psi_1\psi_2 \rangle \quad \text{(Eq. 10)}$$
$$= I_1 + I_1 + 2 \times 0.6I_1 = 3.2I_1.$$

Contrast that case with what happens at the first dark fringe. The waves at that point are shown in Figure 34(c), and the graphs show that ψ_2 is delayed by $0.5T_{av}$ relative to ψ_1. With this delay, the two waves are out of phase for most of the time. Thus $\psi_1\psi_2$ is almost always negative, and for the graphs sketched in Figure 34(c), $\langle \psi_1\psi_2 \rangle \approx -0.9I_1$. The resultant intensity is therefore

$$I_r = I_1 + I_1 + 2 \times (-0.9I_1) = 0.2I_1.$$

☐ How do the intensities at this first dark fringe and at the first bright fringe compare with the intensities of the corresponding fringes produced with monochromatic light?

■ With the broad bandwidth light, the intensities are $0.2I_1$ and $3.2I_1$, whereas for monochromatic light the intensities would be zero and $4I_1$.

The causes of these differences are the phase changes and amplitude changes in the waveform of the illumination from the broadband source. These changes mean that the waves are never exactly out of phase at all times when they arrive at the dark fringe, nor are they exactly in phase at all times at the first bright fringe. Thus complete destructive interference ($I_r = 0$) does not occur, and nor does complete constructive interference ($I_r = 4I_1$).

The other two parts of Figure 34, (e) and (f), show the waves at positions in the diffraction plane where the time shifts between the waves are $4T_{av}$ and $4.5T_{av}$. In both of these cases, $\langle \psi_1\psi_2 \rangle \ll I_1$ and so $I_r \approx 2I_1$. Thus fringes are *not* observed at these two positions. In contrast, with monochromatic light, a dark ($I_r = 0$) fringe would be observed where the time shift was $4.5T_{av}$ and a bright fringe ($I_r = 4I_1$) where the shift was $4T_{av}$.

Why have the fringes disappeared? It is because the phase changes in the waves that take place in an interval of $4T_{av}$ are sufficiently large that there is essentially no correlation — no coherence, if you like — between the phase of the wave at one instant and the phase at a time $4T_{av}$ later. This lack of phase correlation means that $\langle \psi_1\psi_2 \rangle$ is zero, so interference does not occur. Thus the disappearance of the fringes marks the disappearance of the phase coherence between the two waves. There are only about two fringes visible on each side of the central fringe in Figure 34(a), and this indicates that the phase correlation of the illumination disappears for time differences greater than about $2T_{av}$.

5.4 Quantifying temporal coherence: coherence time and coherence length

The discussion in the previous sections has indicated that a laser beam has high temporal coherence and white light from a tungsten filament lamp has low temporal coherence. But it would be useful if we could quantify the degree of temporal coherence of the illumination, rather than just using vague adjectives like 'high' and 'low'. How can this be done?

One way that temporal coherence is quantified is by quoting a value for the so-called **coherence time** τ_c of the illumination. The coherence time is the time interval over which there is some correlation in the phase of a wave. For a perfect monochromatic wave, the coherence time would be infinite, since the sinusoidal pattern would have infinite extent. Knowing the phase of the wave at a particular point at any time, we could predict the phase at that point at any subsequent time. In contrast, for white light, the frequent phase changes that occur in the waveform mean that the coherence time is very short, perhaps only a few periods, or less than 10^{-14} seconds.

The double slit provides one way of making a measurement of the coherence time of illumination. Remember, the light arriving at the centre of the diffraction pattern formed by a double slit takes the same time to travel from each slit, but light arriving at off-axis points takes different times to travel from the two slits. The further from the centre of the diffraction pattern we go, the greater this difference in the time to travel from the two slits. So to measure the coherence time, we observe where the fringes disappear from the diffraction pattern, and this indicates where the phase coherence between light from the two slits disappears. The coherence time is then simply the difference in the times required to travel from the two slits to the point of fringe disappearance.

Fortunately, no geometry or trigonometry is necessary to calculate this time difference. For the first bright fringe, there is one extra wavelength in the path travelled by light from one slit; for the nth bright fringe, there are n extra wavelengths in one path. Each extra wavelength corresponds to a travel time of λ_{av}/c, which is equal to the average period T_{av} of the wave. So if we observe n bright fringes each side of the centre of the diffraction pattern before the fringes disappear, we can say that the coherence time is nT_{av}.

Earlier, in Section 5.2, I asserted that there is a reciprocal relationship between temporal coherence and bandwidth. Large bandwidth means low temporal coherence, and narrow bandwidth means high temporal coherence. Now that I have introduced the coherence time τ_c as a measure of temporal coherence, I can quantify this reciprocal relationship in the following way:

$$\tau_c \approx 1/\Delta f, \qquad (19)$$

where Δf is the bandwidth expressed as a frequency spread.

There is a third way in which the temporal coherence is often quantified, and, perhaps surprisingly, it involves quoting a length. This is known as the **coherence length** l_c. The coherence length is the distance travelled by light in the coherence time, that is

$$l_c = c\tau_c. \qquad (20)$$

This is the distance between two points, *in the direction that the light is travelling*, over which there is a correlation between the phase of the light wave. The coherence length is a particularly useful way to quantify the temporal coherence of illumination because it specifies directly the path difference between two beams (derived from the same source) that should not be exceeded if the beams are to interfere when superposed. As you will see later in the Course, the coherence length of laser illumination is a vital factor that has to be considered in the design of a holography system.

With a double slit, we observe clear fringes at points in the diffraction pattern where the path difference from the two slits is much less than the coherence length, and the fringes disappear when the path difference becomes equal to the coherence length.

> ☐ If five bright fringes are observed on each side of the diffraction pattern produced by a double slit when it is illuminated by light with average wavelength 500 nm, what is the coherence length of the light?
>
> ■ The coherence length is about 5×500 nm, or $2.5\,\mu$m. Remember, the nth bright fringe corresponds to a path difference of $n\lambda_{av}$. Fringes are observed, indicating some phase coherence, for path differences up to five wavelengths, but not for larger path differences. This means that the coherence length is about $5\lambda_{av}$.

Estimating the coherence length of illumination by using a double slit is not very satisfactory. There are two major limitations to the method. First, the maximum coherence length that can be measured is equal to the slit spacing d, since this is the maximum path difference between light from the two slits, and occurs when $\theta = 90°$. This means that double slits are not used for measuring the coherence length of lasers, which may be several centimetres or even metres.

The second limitation applies even to measurements of much shorter coherence lengths, and arises because of the finite width of each of the double slits. As shown in the answer to ITQ 12, the top-hat function that represents the slit width will give rise to a sinc-squared variation of the fringe intensity, in addition to the dependence due to the finite coherence length. This makes it difficult to estimate the coherence length from the diffraction pattern.

Fortunately, there is an alternative way to measure coherence lengths. This uses an instrument called a Michelson interferometer, and this is the subject of the next section.

5.5 The Michelson interferometer

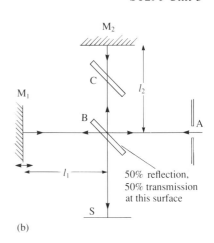

Figure 35 Schematic diagram of the components of a Michelson interferometer.

The **Michelson interferometer** was invented by A. A. Michelson in 1881, and one of its early and famous uses was to investigate (and disprove!) the hypothesis that light propagated in a pervasive medium — the so-called ether — that filled all of space. Nowadays, however, it is widely used for measuring coherence lengths and investigating spectra. Figure 35 shows schematically the basic components of a Michelson interferometer.

A beam of light enters the instrument via aperture A and is divided by a beam-splitter B into two beams, each with half of the original intensity. One beam is transmitted through the beam-splitter and this beam travels to the mirror M_1. It is reflected back to B, where it is split again, half the intensity being reflected to the screen S, the remainder passing out through A. The remainder of the light entering at A is reflected by the partially silvered front face of B. This second beam travels to the mirror M_2 via a transparent compensating plate C. The beam is then reflected back to B where it is split again, half the intensity being transmitted to the screen S, the rest being reflected out through A. The compensating plate ensures that the optical paths each beam traverses are equivalent. It compensates for the fact that the first beam travels through the glass of the beam-splitter three times before reaching the screen S, whereas the second beam only travels through it once.

The net field on the screen S is therefore due to a wave that has travelled there via M_1 and an identical wave that has travelled via M_2. The two waves will arrive in step with each other *if* the optical distances l_1, l_2 (between the mirrors M_1, M_2 and the beam-splitter) are the same. But one wave can be delayed with respect to the other by making the lengths l_1 and l_2 different. The path difference Δl is $2(l_1 - l_2)$ and the corresponding time difference Δt is $2(l_1 - l_2)/c$. We thus have a means of comparing a wave with itself at an adjustable time delay Δt. By moving one of the mirrors (M_1 in Figure 35) and observing the intensity on the screen, we will know when Δt approaches the coherence time τ_c because we will no longer be able to observe interference. In this instrument, interference manifests itself via an oscillation of the intensity at the screen as M_1 is moved. When $\Delta l = 0, \pm \lambda_{av}, \pm 2\lambda_{av}$, etc., the intensity will be large since the two waves arrive exactly in phase (constructive interference). Conversely, when $\Delta l = \lambda_{av}/2, 3\lambda_{av}/2$, etc., the intensity will be small since the waves arrive exactly out of phase (destructive interference). The intensity will be constant at an intermediate level when the waves are no longer correlated; this happens when Δt exceeds the coherence time τ_c which means that the path length Δl is longer than the coherence length l_c. Thus the path difference $2(l_1 - l_2)$ at which the intensity on the screen no longer varies as the mirror is moved is a measure of the coherence length, and the coherence time and the bandwidth Δf can be calculated from this by using the relationships in equations 19 and 20.

But we can do more than just obtain the bandwidth Δf. We can obtain detailed information about the spectrum by means of a technique called **Fourier transform spectrometry**, which involves observing the precise manner in which interference effects disappear in a device such as a Michelson interferometer. This is the subject of the Video Programme 5.

You should now watch Video Programme 5 'Michelson interferometer'. This programme reinforces the concept of temporal coherence, and the relationship between coherence length, coherence time and bandwidth. It also shows how the Michelson interferometer is used, both in a student laboratory and also in an industrial laboratory for determining the composition of the exhaust gases from a car engine. You will find notes about the programme in the Video Notes booklet.

5.6 Summary of Section 5

1 The illumination of an object can be characterized by its *temporal coherence* (related to the bandwidth of the illumination) and by its *spatial coherence* (related to the range of inclinations of the illumination).

2 There is a reciprocal relationship between temporal coherence and spectral bandwidth; narrow bandwidth leads to high temporal coherence, and wide bandwidth leads to low temporal coherence.

3 The temporal coherence of illumination can be quantified by quoting:

(i) the *coherence time* τ_c, which is the time over which there is some correlation in the phase of the wave;

(ii) the *coherence length* l_c, which is the length of wave train over which there is some correlation in the phase of the wave;

(iii) the temporal frequency *bandwidth* Δf, which is the range of temporal frequencies present in the illumination.

These three quantities are related by the following equations:

$$l_c = c\tau_c \qquad \text{(Eq. 20)}$$

$$\tau_c = 1/\Delta f. \qquad \text{(Eq. 19)}$$

4 A double slit can be used to estimate the temporal coherence of illumination. If there are n bright fringes in the double-slit diffraction pattern between the central bright fringe and the region where the fringes disappear, then

$$\tau_c \approx nT_{av}, \quad \text{and} \quad l_c \approx n\lambda_{av}.$$

This method only works if the coherence length is less than the slit separation.

5 According to the Fourier model, the disappearance of the fringes is explained by the fact that each wavelength in the illumination produces a diffraction pattern with a different fringe spacing, and the fringes of these different patterns get out of step away from the centre of the pattern.

6 According to the Huygens model, the disappearance of the fringes is explained by lack of phase coherence of the waves from the two slits when the difference in path lengths between the slits and the diffraction plane is greater than the coherence length (or equivalently, when the time delay between the two waves is greater than the coherence time).

7 A Michelson interferometer can be used to measure the coherence length of illumination. The spectrum of the illumination used in a Michelson interferometer can be computed by taking the Fourier transform of the interferogram, i.e. the intensity versus mirror position. This technique is known as Fourier transform spectrometry, and it is widely used for chemical analysis.

SAQ 8 The bandwidth, coherence time and coherence length of the visible illumination from a tungsten lamp can be estimated from the measurements that you made (in Section 2.5.1) on the diffraction pattern produced by a grating.

(a) Use your results from ITQ 3(c) to estimate the bandwidth $\Delta\lambda$ of the visible illumination from the tungsten lamp.

(b) Hence estimate the bandwidth Δf, the coherence time τ_c, and the coherence length l_c of the illumination.

SAQ 9 (a) Suppose that the tungsten lamp in the Home Kit was used to illuminate a double slit. Use the answers to the previous SAQ to estimate the number of bright fringes that you would expect to be detected on either side of the central bright fringe. (You should assume that the lamp acts as a point source.)

(b) The same lamp, again with a point source aperture, is used in the two mirror system shown in Figure 22(a). The distance between the two mirrors is 0.1 m. Would you expect to observe interference fringes in the overlap region P in Figure 22(a)? Explain the reason for your answer.

(c) Would you expect to observe interference fringes if an expanded laser beam with a coherence length of 0.5 m were used in place of the point source and convex lens in Figure 22(a)?

SAQ 10 A Michelson interferometer (Figure 35) is used to measure the bandwidth of a helium–neon laser. Mirror M_1 is moved back (starting from the position where the path lengths of the two beams are equal), and the interference at screen S ceases when the mirror has moved 0.15 m. Calculate (a) the coherence length, (b) the coherence time, and (c) the bandwidths, Δf and $\Delta\lambda$, for this laser.

6 Spatial coherence

The **spatial coherence** of the illumination at an object is determined by various geometrical factors: the size of the source, the shape of the source, the distance between the source and the object, and the effects of any collimating lenses or apertures between the source and the object. We were able to ignore these factors in the previous section by only considering illumination from a very distant point source. In the region of the object, such illumination was regarded as a plane wave travelling in a single direction. This simplification allowed us to concentrate on how the source's bandwidth determines the *temporal* coherence of the illumination.

In this section, we will discuss how the size and shape of the source affect the *spatial* coherence of the illumination at the object. This will give a theoretical basis to Experiment 7, in which you observed the effects of opening the iris that controlled the source size. Because we will be concerned here with *spatial* coherence, we will assume that the illumination has a narrow spectral bandwidth, so that the effects of temporal incoherence can be ignored.

Now, in general, a complete description of the illumination requires knowledge of the source–object distance, because this distance determines the curvature of the wavefronts at the object. However, to simplify the discussion, we will only consider situations where the illumination from each point on the source arrives at the object as a plane wave. Two ways in which such illumination can be produced are shown in Figure 36. The source can be placed a large distance from the object, or it can be placed in the *front* focal plane of a converging lens.

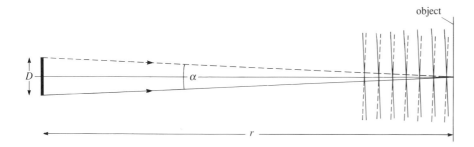

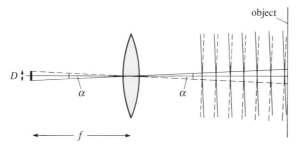

Figure 36 Two sources that produce plane wave illumination with a range of inclinations. Plane waves from the top and bottom of each source are shown arriving at the object. (a) Distant source; range of plane wave inclinations $\alpha = D/r$. (b) Source in front focal plane of lens; $\alpha = D/f$.

In both cases, each source point produces a plane wave, the inclination of which is determined by the position of the point on the source. The angular range of plane wave inclinations in the illumination is equal to the angular size α of the source, and this is related to the source size D by the following relations:

$$\text{source at (large) distance } r: \quad \alpha = \frac{D}{r}; \tag{21}$$

$$\text{source in focal plane of lens:} \quad \alpha = \frac{D}{f}. \tag{22}$$

The angular size α of the source is the most important factor determining the spatial coherence of the illumination. When α is very small, or, in other words, when the source is effectively a point, then the illumination is spatially coherent. As the angular size of the source increases, the illumination becomes less coherent, and eventually it becomes spatially incoherent. To illustrate and explain the effects of source size, I will consider the diffraction pattern of a double slit object, and I will show that the Fourier and Huygens approaches again provide complementary insights into spatial coherence.

6.1 The Fourier approach to spatial coherence

By now, you should be very familiar with the diffraction pattern produced when a double slit is illuminated by a very distant point source. If the point source is on the optical axis, then the illumination is a plane wave travelling parallel to the optical axis, and the diffraction pattern is a cosinusoidal intensity distribution, with the central bright fringe on the optical axis (Figure 37(a)). When the point source is displaced from the optical axis by an angle δ (Figure 37(b)), the illumination is a plane wave travelling at an angle δ to the optical axis; in this case, the central bright fringe of the diffraction pattern is displaced by the same angle δ from the optical axis. Suppose though that we use an extended source: what sort of diffraction pattern would be produced in this case?

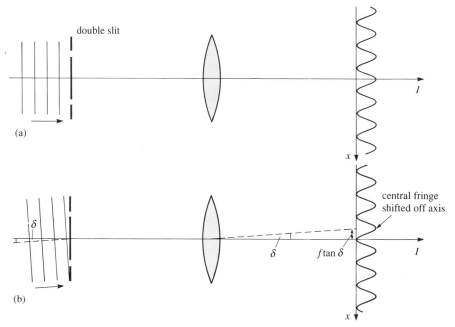

Figure 37 Double slits illuminated by plane waves from distant point sources (a) on the optical axis and (b) displaced from the axis by angle δ. In both cases, the diffraction pattern in the back focal plane of the lens is a cosinusoidal intensity distribution, but the central bright fringe is on the axis in (a) and displaced by angle δ from the axis in (b).

Well, we can regard an extended source as made up of many closely spaced point sources. Each of the point sources will produce a cosinusoidal intensity distribution in the diffraction plane, but the displacement of each cosinusoidal distribution from the optical axis will be different. Since each source point emits waves completely independently of other points — that is, the waves from different points are uncorrelated — the resultant intensity in the diffraction plane will simply be the sum of all of the individual displaced cosinusoidal intensity distributions.

Figure 38 illustrates the effect of adding the cosinusoidal double-slit diffraction patterns produced by source points on three extended sources. Each of the sources is a line (like a straight filament) and is oriented perpendicular to the optical axis and perpendicular to the double slit. For the shortest line source (part (a) of the figure), the relative displacements of the different cosinusoidal fringe patterns are small compared with the period of the fringes. At first sight the resultant intensity, shown at the bottom of the figure, may appear similar to the individual intensity distributions. However, though the fringe spacing in the resultant pattern is the same as the fringe spacing in the individual patterns, if you look closely you will see that the resultant

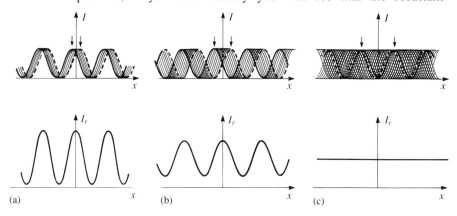

Figure 38 Double-slit diffraction patterns produced with illumination from three line sources; the angular size of the source is smallest in (a), and largest in (c). *Top*: diffraction patterns produced by illumination from individual points on the line sources; the cosinusoidal patterns produced by points at the two ends of the source are shown by a bold continuous line and a bold broken line, and the central fringes produced by these points are indicated by arrows. *Bottom*: resultant intensities due to the extended line sources. (The patterns are scaled so that the average intensity is the same size in each graph.)

intensity never drops to zero. This is because the zero intensity positions — the dark fringes — don't quite coincide. Also the maximum intensity never quite reaches twice the average intensity because the individual peaks don't coincide.

For the source with the intermediate length, the relative displacement of the fringes (shown at the top of Figure 38(b)) is about half of the fringe spacing. The resultant intensity is still cosinusoidal, and still has the same fringe spacing as in the individual patterns. However, the displacements of the individual patterns produced by different points on the source lead to a larger minimum intensity, and smaller maximum intensity, relative to the average intensity.

The length of the longest source is such that the fringes produced by one end of the source exactly coincide with the fringes from the opposite end. However, the *central* fringes are separated by one fringe spacing, and in this special situation the bright fringes are uniformly distributed across the diffraction plane. The resultant intensity is therefore the same at all points, and this means that no fringes are visible in the diffraction plane.

Now the difference between the three resultant intensity graphs at the bottom of Figure 38 can best be specified in terms of a quantity known as the **modulation**, which is defined in Figure 39. If I_{max} and I_{min} are the maximum and minimum intensities in the diffraction pattern, then the modulation M of the pattern is given by the equation

$$M = \frac{I_{max} - I_{min}}{I_{max} + I_{min}}. \qquad (23)$$

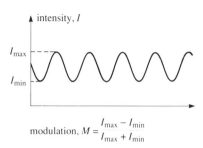

Figure 39 The definition of modulation.

From this definition it is clear that when $I_{min} = 0$, then the modulation is $M = 1$; this is the largest possible value of the modulation. It corresponds to the maximum possible contrast between bright and dark fringes, so the fringes have the highest visibility. Conversely, when $I_{min} = I_{max}$, the modulation M is *zero*; in this case, no fringes are visible in the diffraction pattern. It is worth noting that, because the modulation of the intensity determines how visible the fringes are, the term **visibility** is often used in place of modulation to describe the quantity defined by equation 23.

> **ITQ 15** What are the modulations of:
>
> (a) the diffraction patterns produced by individual source points, shown at the top of Figure 38?
>
> (b) the resultant diffraction patterns shown at the bottom of Figure 38(a), (b) and (c)?

Now you should remember from the discussion of temporal coherence in Section 5 that the *number* of fringes observed in the diffraction pattern of a double slit provides an indication of the *temporal* coherence of the illumination. If a large number of fringes are observed, then the source has high temporal coherence, and if only a few fringes are observed, the illumination has low temporal coherence (Figure 29). It turns out that the fringes in the diffraction pattern of a double slit also provide information about the *spatial* coherence of the illumination. However, it is the *modulation* of the diffraction pattern that is related to the spatial coherence, not the number of fringes. This relationship is summarized in Figure 40. High modulation, $M \approx 1$, indicates

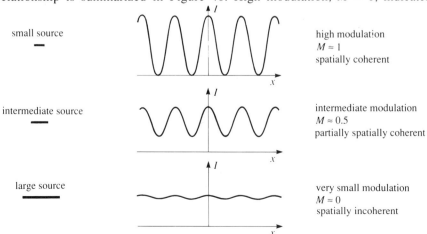

Figure 40 Diffraction patterns of a double slit produced by extended sources of three different sizes. As source size increases, modulation decreases and spatial coherence decreases.

that the illumination is spatially coherent (Figure 40(a)), and low modulation, $M \approx 0$, indicates that the illumination is spatially incoherent (Figure 40(c)). These are the two extreme cases. The diffraction pattern in Figure 40(b) has a modulation of 0.5: in this case, the illumination of the double slit is said to have partial spatial coherence, and the degree of partial spatial coherence is numerically equal to the value of the modulation of the diffraction pattern.

6.2 The extent of spatial coherence

The discussion so far has indicated that the modulation of the diffraction pattern of a double slit is lower when an extended source is used than when a point source is used. In addition, the reduction in modulation has been linked with a reduction of the spatial coherence of the illumination at the slits. I now want to quantify the relationship between source size and spatial coherence. In particular, I want to answer the question 'How large can the source be made before the illumination at the slits is spatially incoherent?'

Have another look at Figure 38. As the source size increases from left to right of the figure, the modulation of the resultant intensity decreases. The modulation is zero in Figure 38(c), so the illumination at the two slits has zero spatial coherence. Note that the central fringes produced by the two ends of the source are displaced by exactly one fringe spacing in this case.

Now if the distant source has an angular size α when viewed from the object, the angular separation between the centres of the fringe patterns produced by the top of the source and by the bottom of the source will then also be α (Figure 41). But for the zero modulation case in Figure 38(c), this separation α is equal to the angular separation between neighbouring fringes in the fringe pattern.

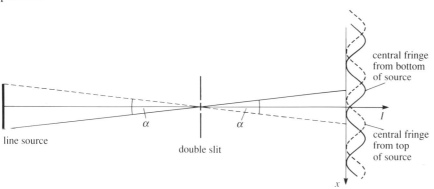

Figure 41 The angular shift between diffraction patterns from top and bottom of the source is equal to the angular size of the source.

For a source point on the optical axis, the bright fringes occur at angles θ_n that satisfy the expression derived in ITQ 14:

$$\sin \theta_n = n\lambda_{av}/d, \qquad (24)$$

where d is the slit spacing, λ_{av} is the average wavelength of the narrow-bandwidth illumination, and n is an integer. Assuming that the angles are small, so that we can make the approximation $\sin \theta_n \approx \theta_n$, then the angle $\Delta\theta$ between neighbouring fringes (e.g. between $n = 0$ and $n = 1$) is simply

$$\Delta\theta = \lambda_{av}/d.$$

For zero modulation, $\alpha = \Delta\theta$, and so

$$\alpha = \lambda_{av}/d. \qquad (25)$$

Now, this equation can be interpreted in two, complementary ways. First, suppose that we wish to produce a diffraction pattern of a double slit with spacing d. Equation 25 then tells us the minimum angular size α of a line source that will produce zero modulation in the diffraction pattern. For this value of α, we will observe no fringes, and no information about the slits can be obtained from the diffraction pattern: the illumination at the slits is said to be *spatially incoherent*. For smaller values of α, though, that is for $\alpha < \lambda_{av}/d$, fringes *will* be observed in the diffraction pattern, and the illumination of the slits is said to be *partially spatially coherent*. The condition for a clear well-modulated diffraction pattern is that $\alpha \ll \lambda_{av}/d$.

The second way of looking at equation 25 is to regard it as a way of defining the distance in the object plane over which the illumination from a given line source is coherent. Rearranging equation 25 gives

$$d = \lambda_{av}/\alpha. \tag{26}$$

For a line source of angular size α, this equation then specifies the slit separation d at which no fringes are observed in the diffraction pattern. For smaller slit separations than specified by equation 26, i.e. for $d < \lambda_{av}/\alpha$, fringes will be observed.

☐ Why does reducing the slit spacing d, from an initial value of $d = \lambda_{av}/\alpha$, lead to fringes appearing in the diffraction pattern?

■ When $d = \lambda_{av}/\alpha$, the angular spacing $\Delta\theta$ of the fringes is equal to the angular size of the source, as shown in Figure 38(c), and the fringes produced by the individual points on the source cancel each other. Reducing the slit spacing increases the fringe spacing so that it is greater than the angular size of the source. The central fringes from individual points on the source then no longer cover a distance equal to the (increased) fringe spacing, and so the fringes only partially cancel each other.

The value of d given by equation 26 is called the **coherence width** of the illumination, and we will represent it by the symbol w_c. Thus

$$\text{coherence width (line source):} \quad w_c = \lambda_{av}/\alpha, \tag{27}$$

and this quantity tells us the distance *in the object plane* over which spatial coherence of the illumination exists. If two points on an object are separated by a distance that is very much less than the coherence width, then the illumination of those points will be spatially coherent. However, as the separation increases, the spatial coherence decreases, and when the separation is w_c the illumination of the two points is spatially incoherent.

It is important to distinguish clearly between coherence *width* w_c, which is a measure of *spatial* coherence, and coherence *length* l_c, which was introduced in Section 5.4 as a measure of *temporal* coherence. The coherence width is the distance, perpendicular to the direction in which the light travels, over which there is a correlation between the field at two points. The coherence length is the distance in the direction in which the light travels over which there is a correlation between the field at two points. Spatial coherence and the coherence width depend on the angular size and shape of the source, since these factors determine the range of inclinations of the illumination. Temporal coherence and coherence length depend on the bandwidth of the illumination.

It is also important to note that the expression for the coherence width in equation 27 was derived for a *line* source orientated perpendicular to the double slit (see Figure 41). It defines a coherence width in the object plane in the direction *parallel* to the length of the source. The coherence width in a direction *perpendicular* to the line source is very much larger, because the angular size of the source in this direction is very much smaller. The expression in equation 27 is still valid, but in this case α is the angular size corresponding to the narrow dimension of the source.

Perhaps the most important sources in practice are circular, and in this case the expression for the coherence width is slightly different from equation 27. For a source, with angular *diameter* α viewed from the object,

$$\text{coherence width (circular source):} \quad w_c = 1.22\lambda_{av}/\alpha. \tag{28}$$

The factor of 1.22 in this expression is identical to the numerical factor in the expression for the radius of the Airy disc — the central region in the diffraction pattern of a circular aperture — which was introduced in Units 3 & 4, and the reason for this identity will become clear in Section 6.4.

The fact that this coherence width is larger than for a line (or rectangular) source of angular length α is easy to explain qualitatively. Suppose the circular source is divided into strips parallel to the double slit, as shown in Figure 42. Then each strip will produce a high-modulation cosinusoidal fringe pattern. However, the strips at the edge will produce lower intensity fringes than the strips at the centre, because the edge strips are shorter. The edge strips are therefore less effective at cancelling the fringes produced by the central strips than if all strips were of equal length; this means that fringe modulation will still be observable when the slit spacing d is λ_{av}/α. The slit spacing can be up

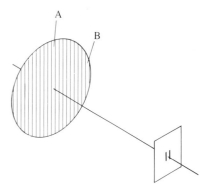

Figure 42 A circular source divided into strips parallel to a double slit that it is illuminating. Strip A produces brighter fringes than strip B, because A is longer than B.

to 22% larger (so that the fringe spacing is smaller) before the fringe modulation falls to zero, and this accounts for the factor of 1.22 in the expression for the coherence width of a circular source in equation 28.

Now this factor of 1.22 may seem fairly trivial, and it is probably true to say that a 20% change in coherence width is not usually very significant in optical systems. However, it does indicate that spatial coherence depends on the shape of the source, and this has very important applications. Indeed, you will see in Section 6.4 how measurements of the variation of fringe visibility (i.e. the variation of the degree of spatial coherence) with slit spacing can be used to deduce information about the size, shape and intensity distribution in a source.

6.3 The Huygens approach to spatial coherence

I now want to show how the Huygens model can give further insight into the concept of spatial coherence.

When discussing double slit diffraction using this model, we have to consider interference of the waves that passed through the two slits. We therefore need some information about the relationship, or the correlation, between the waves at these slits. Let's take the simplest case first: a double slit illuminated by a distant point source on the optical axis, as shown in Figure 43(a). (We will again assume that the spectral bandwidth of the light is narrow, so that the illumination has high temporal coherence.) In this case, the fields at the two slits have exactly the same time dependence, since the wavefronts from the point source reach them simultaneously. Thus the two slits act as perfectly correlated secondary sources, and this means that superposition of the waves from the slits will result in complete constructive interference at some points in the diffraction plane and complete destructive interference at other points. The modulation of the cosinusoidal diffraction pattern will therefore be $M = 1$, and the illumination of the slits is said to be spatially coherent. Remember that this illumination is also temporally coherent, because we have assumed that the spectral bandwidth is narrow, and so this situation is identical to that discussed at the start of Section 5.3.2.

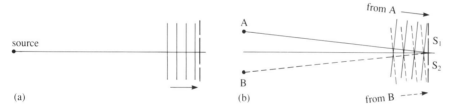

Figure 43 (a) Distant point source producing plane wave illumination at a double slit; the wavefronts arrive simultaneously at the two slits. (b) Two point sources producing two inclined plane waves; wavefronts from A reach S_1 before S_2 but wavefronts from B reach S_1 after S_2.

Now consider what happens when the illumination comes from *two* distant point sources, each producing temporally coherent light with the same average wavelength. This situation is shown in Figure 43(b). Here we have two plane waves, with different inclinations, illuminating the double slit, and in this case the resultant fields at the the two slits are *not identical*. The reason for the difference lies in the fact that the wavefronts from point source A reach the top slit S_1 *before* they reach the bottom slit S_2 and, conversely, the wavefronts from point source B reach slit S_1 *after* they reach slit S_2. So although the resultant wave at each slit is a superposition of waves from the same two point sources, the two resultant waves are different because the component waves are added with different phase shifts — different time delays — at the two slits. The consequence is that the two slits do *not* act as identical secondary sources; they are not perfectly correlated.

The reduction in the correlation between the waves from the two slits means that complete constructive interference and complete destructive interference are no longer possible. Thus the modulation of the diffraction pattern will be less than one, which indicates that the illumination is only partially spatially coherent.

Now a similar argument can be applied to illumination from an extended source, since this can be thought of as a large number of point sources that produce plane waves with a range of inclinations. As the size of the source increases, the relative phase shifts of the waves reaching the two slits become larger, and this means that the correlation between the resultant fields at the two slits decreases, and the spatial coherence decreases.

The Huygens model also gives an insight into the gradual reduction in modulation of the diffraction patterns of double slits when the spacing of the slits is increased. Suppose that we have a temporally coherent extended source of a particular size. Then if the slits are extremely close together, there will only be very small phase differences between the fields at the two slits arising from any particular point on the source, as shown in Figure 44(a). This means that the resultant fields at the two slits will be very similar, so these fields will be highly correlated, and the diffraction pattern will have a high modulation. However, as the separation of the slits increases, the phase differences between the waves arriving at the two slits will increase (Figure 44(b)). The correlation between the resultant fields at the two slits will therefore become smaller, and so the modulation of the diffraction pattern will be reduced. Increasing the slit spacing will eventually lead to a situation where there is no correlation between the resultant fields at the two slits, and then no fringes will be observed in the diffraction pattern. The smallest slit separation at which this occurs corresponds to the coherence width w_c of the illumination.

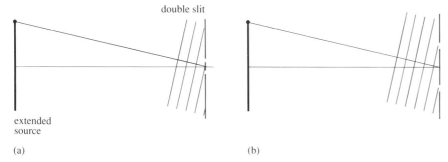

Figure 44 Extended source illuminating double slits with two spacings. (a) Very small slit spacing; here the phase difference of the waves from the top of the source at the two slits is much less than a wavelength (i.e. the spacing of the wavefronts). (b) Large slit spacing; here the phase difference at the two slits is about 2π.

So according to the Huygens model, we can think of spatial coherence in the following way. For a small source, the wave fields at two spatially separated locations on the object (e.g. at two slits) are very similar. Because the two fields are highly correlated, constructive and destructive interference will be observed in the diffraction pattern. However, with a large source, the resultant fields at the two slits are different, because waves from different regions of the source add together with different phase shifts at each of the slits. The larger the source, or the larger the spacing of the two slits, the smaller will be the correlation between the fields at these two slits, so the smaller will be the spatial coherence of the illumination and the smaller will be the modulation in the diffraction pattern. Note that the basic waveforms of the field at the two slits will have the same average frequency, and the fluctuations in amplitude and phase will have similar magnitudes, *irrespective of the size of the source*. But as the source is made larger, the correlation between the changes in amplitude and phase at the two slits becomes smaller, which means that the spatial coherence of the illumination is reduced.

6.4 Investigating the structure of self-luminous objects

When discussing the extent of spatial coherence (Section 6.2), I stated that, for a circular source, the visibility of the fringes in the diffraction pattern of a double slit is zero when

$$d = w_c = 1.22 \lambda_{av}/\alpha \qquad \text{(Eq. 28)}$$

In this equation, d is the spacing of the slits, w_c is the coherence width of the illumination, λ_{av} is the mean wavelength of the source and α is its angular diameter. This result is the basis of a technique for measuring the size of stars.

The technique makes use of an instrument called a **Michelson stellar interferometer**, which was invented by A. A. Michelson in 1920, some 40 years after he invented the interferometer described in Section 5.5. The device is shown in Figure 45. Essentially it provides a way to observe double slit diffraction patterns created with illumination from a star. Circular mirrors M_1 and M_2 act as the two slits*, and their separation d can be adjusted by moving them closer together or further apart. The other mirrors and the objective lens

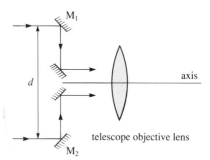

Figure 45 The Michelson stellar interferometer.

---

* The change from a long slit to a circular mirror has no significant effect, as you can see from plates 1.2 and 1.3 of the Atlas. The diffraction patterns of two small circular apertures is dominated by 'double slit' interference fringes, the spacing of which is inversely proportional to the separation of the apertures. The circularity of the apertures is shown by the overall Airy ring pattern envelope of the diffraction pattern.

of the telescope serve only to display the diffraction pattern in a convenient manner.

The procedure is to increase the separation d of mirrors M_1 and M_2 until the fringes produced by a star disappear. From the value of d at which the fringes disappear it is possible to calculate the angular diameter α of the star. To determine the same stellar diameter with a conventional telescope the objective lens would need to have a *diameter* equal to d. However, the cost and difficulty of making lenses increases very rapidly with their diameter. It is therefore cheaper to use a smallish diameter telescope fitted with a Michelson stellar interferometer rather than attempt to build a very large conventional telescope. Furthermore, a telescope's performance is not only limited by its size, but also by atmospheric turbulence. Therefore a telescope's *effective* diameter — the notional diameter associated with its ability to resolve spatial detail — is considerably less than its actual diameter. Michelson's stellar interferometer was less affected by atmospheric turbulence and this even further increased its advantage.

Now, if we assume that the star is effectively a uniform disc of angular diameter α, then the critical separation of the two mirrors at which the fringes disappear is the coherence width w_c, which is given by

$$w_c = 1.22 \lambda_{av} / \alpha. \qquad (Eq.\ 28)$$

Thus measurements of the critical mirror spacing w_c and the mean wavelength of the light from the star allow the angular diameter α to be determined.

The star Betelgeuse (pronounced 'beetle-juice') in Orion was the first to have its diameter measured by the Michelson stellar interferometer. The device was used in conjunction with the 100-inch diameter reflecting telescope at Mt Wilson, California, which was the largest telescope in the world at that time. The interferometer could work out to a mirror separation of 240 inches, which is almost $2\frac{1}{2}$ times larger than the telescope's diameter.

> **ITQ 16** (a) On a December night in 1920, the fringes produced by Betelgeuse vanished at a mirror separation of 121 inches (3.07 m). Assuming that the mean wavelength of the light from this star is 570 nm, calculate the angular diameter of Betelgeuse.
>
> (b) What is the smallest stellar diameter that Michelson's first stellar interferometer could have measured?

It turns out that Betelgeuse is a very large and comparatively close star. There are only a few other stars with angular diameters large enough to be measured with a 240-inch mirror separation, and attempts to increase the separation beyond 240 inches ran into practical difficulties. However, more recently, the same interferometer principle has been applied to radiotelescopes, and sources with extremely small angular diameter have had their angular diameters successfully measured.

But at this point a note of caution must be sounded. Equations 27 and 28 both give conditions for fringes to disappear. They are different because the spatial form of the source is different in each case — a uniform line (or rectangle) for equation 27 to apply, and a uniform disc for equation 28. In order to apply the correct equation, we have to know something about the spatial form of the source. The mere observation of the value of d at which fringes disappear does not unambiguously tell us the angular size of the source. However, this ambiguity can be removed by observing how the fringe modulation changes as the slit spacing is changed.

In principle, such measurements are straightforward. The telescope, with mirrors mounted on it, is aimed at the star(s) in question. With the two mirrors that form the double-slit positioned close together, a measurement is made of the modulation at the centre of the diffraction pattern. The mirror spacing is increased, and the modulation measured again. This procedure is then repeated until fringes can no longer be detected in the diffraction pattern. Figure 46 shows the result that might be obtained when investigating a single circular star.

Had the source been rectangular (perhaps unlikely for a star!) with one edge parallel to the line separating the mirrors, then the modulation versus separation graph would have had the form shown in Figure 47, which has a

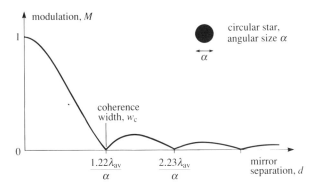
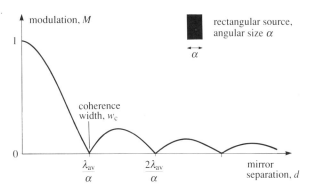

Figure 46 Fringe modulation versus mirror separation obtained when a stellar interferometer is used to investigate a single circular star.

Figure 47 The fringe modulation versus mirror separation for a rectangular source (assuming that the mirror separation is parallel to one edge of the rectangle).

different shape to the graph shown in Figure 46. Note that the smallest separation at which the modulation is zero is different in the two cases, and these separations correspond to the coherence widths defined by equations 27 and 28.

If careful measurements are made of how the fringe modulation changes when the spacing of the mirrors is increased, then it is possible to compute the intensity variation across the source. This technique is widely used in astronomy, and will be discussed further in the Crab Nebula case study.

6.5 Summary of Section 6

1 The spatial coherence of illumination depends principally on the angular size α of the source. It also depends on the shape of the source.

2 The *modulation M* of an intensity distribution is defined as

$$M = \frac{I_{\max} - I_{\min}}{I_{\max} + I_{\min}},\qquad \text{(Eq. 23)}$$

where $I_{\max}$ and $I_{\min}$ are the maximum and minimum intensities, respectively.

3 A distant point source produces illumination with high spatial coherence; with a double slit object, the modulation of the diffraction pattern produced by such illumination is $M = 1$.

4 As the source size increases, the spatial coherence of the illumination decreases, and consequently the modulation of the double slit diffraction pattern decreases.

5 The *coherence width* w_c is the distance in the object plane over which the illumination is coherent;

$$\text{coherence width (line source):}\quad w_c = \lambda_{av}/\alpha \qquad \text{(Eq. 27)}$$

$$\text{coherence width (circular source):}\quad w_c = 1.22\lambda_{av}/\alpha, \qquad \text{(Eq. 28)}$$

where λ_{av} is the average wavelength and α is the angular length, or angular diameter, of the source.

6 According to the Fourier model, decreasing spatial coherence leads to a reduction in the modulation of the diffraction pattern of a double slit because plane wave components with different inclinations in the illumination produce displaced cosinusoidal fringe patterns.

7 According to the Huygens model, the modulation of the diffraction pattern of a double slit is a measure of the correlation between the resultant waves at the two slits. The reduction in modulation with increasing source size arises because waves from different points on the source have to be superposed with different phase shifts at the two slits.

8 A Michelson stellar interferometer is a double-slit system (on a very large scale) that is used to measure the angular diameters of stars and the intensity variation across them.

SAQ 11 This question is based on the Home Kit optical system shown in Figures 5 and 8, and on results obtained in Experiment 7. In that experiment, you adjusted the aperture diameter so that the neighbouring images of the filament that were observed in the diffraction pattern of the grating were just touching.

(a) What was the range of inclinations α of the plane waves reaching the grating in this situation?

(b) What was the coherence width of the illumination at the grating in the direction parallel to the filament in this situation? (Assume that the yellow–green filter was in place, so that $\lambda_{av} \approx 580$ nm.)

(c) The coherence width that you have calculated should be equal to the spatial period of the grating. Check that this is the case, and explain why the two quantities should be equal in this situation.

SAQ 12 A spherical 'pearl' light bulb with diameter 50 mm, together with a red filter, is used to produce a diffraction pattern of a double slit (slit separation = 10^{-4} m).

(a) If the bulb is placed 10 m from the double slit, what is the coherence width of the illumination at the slits?

(b) Use Figure 46 to estimate the modulation of the diffraction pattern.

(c) Rather than locating the bulb 10 m away, it could be placed 1 m from the double slit and a circular aperture could be placed in front of it to limit its size. What aperture diameter is required if fringes with the same modulation as in part (b) are to be observed with the red filtered light?

SAQ 13 A Michelson stellar interferometer is used to investigate a so-called double star. This is two stars that appear close together in the sky, but with a separation that is much greater than the diameters of the stars; the two stars can therefore be regarded as two point sources. The interferometer is oriented so that the mirror separation is parallel to the line joining the two stars. As the mirror separation is increased, the fringe modulation in the image first falls to zero at a mirror separation of 4.5 m.

(a) Deduce the relationship between the angular separation α of the two stars, the mean wavelength λ_{av} of the starlight and the mirror spacing d at which the fringes disappear. [*Hint* The fringes will disappear when the spacing $\Delta\theta$ of the fringes produced by either star alone is twice the angular separation of the two stars.]

(b) Hence calculate the angular separation of the two stars, assuming $\lambda_{av} = 500$ nm.

(c) If the distance to the double star is known to be 3×10^{19} m, what is the separation between the two stars perpendicular to the line of site?

7 X-ray diffraction

In this concluding section, I want to show an example of the application of the concepts of temporal and spatial coherence to a different type of illumination. For though the experiments and examples in this unit have been concerned with visible light, the concept of coherence is not restricted to this narrow region of the electromagnetic spectrum. In particular, it can be applied to **X-rays**, which are electromagnetic waves with wavelengths of order 0.1 nm, some three or four orders of magnitude shorter than the wavelengths of visible light.

In 1912, Max von Laue discovered that X-rays are diffracted by crystalline solids in much the same way that light is diffracted by gratings. Since that time, the techniques of **X-ray diffraction** have been used to elucidate the internal structure of matter on the atomic scale. Initially, X-ray crystallographers concentrated their efforts on determining the spacing and the arrangement of the atoms in simple crystals. However, half a century after the discovery of X-ray diffraction, it was used to determine the structure of DNA, insulin, haemoglobin and many other complex and important biological molecules.

I will discuss briefly the requirements for the temporal and spatial coherence of the X-ray illumination used in such experiments, but first we must be clear about why X-rays are used for structure determinations.

☐ The spacing d of atoms in solids or in molecules is typically 0.5 nm or less. Explain why light ($\lambda \approx 500$ nm) cannot be used to determine crystal structures.

■ The diffraction condition (equation 1) is $\sin\theta = q\lambda$, or $\sin\theta = \lambda/d$. The *maximum* possible diffraction angle is $\theta = 90°$, corresponding to $\sin\theta = 1$. This means that $\lambda/d \leq 1$, or $d \geq \lambda$. If this condition does not hold, then the transmitted illumination doesn't carry any information about the spacing d. Typical atomic spacings (0.5 nm) are very much less than the wavelengths of light (about 500 nm) and so light can carry no information about the atomic spacings.

X-rays have a much smaller wavelength than light, typically 0.05–0.2 nm. Since this is smaller than atomic spacings, X-rays *are* diffracted by these small-scale structures. However, there are no practical lenses for X-rays, and so in using X-rays as a means of investigating the atomic structure of materials, we have to make do with information from the far-field diffraction pattern rather than a conventional image.

The effects of changes in the temporal and spatial coherence of the X-ray illumination can be seen quite clearly in X-ray diffraction patterns. Consider first the temporal coherence of the illumination. Figure 48 shows the diffraction pattern from a crystal of zinc blende (ZnS). In this crystal, the zinc atoms and the sulphur atoms each form an arrangement that is basically cubic. We will not be concerned here with the relationship between the crystal structure and its diffraction pattern. Suffice it to say that there is an inverse relationship between the spacing of atoms in the crystal and the spacing of spots in the diffraction pattern, just as with optical diffraction patterns. Also the symmetry of the diffraction pattern reflects the symmetry of the crystal structure.

The pattern in Figure 48 was produced using a small source of X-rays that emitted a wide range of wavelengths. You can see that the pattern contains prominent radial streaks, and that each streak exhibits a pair of bright spots. These features are a consequence of the spectrum of the X-rays used to form the diffraction pattern, and this spectrum is shown in Figure 49(a). The spectrum is broad, and it is the continuous range of wavelengths that is responsible for the streaks in the diffraction pattern. Also the spectrum contain two prominent peaks (or rather pairs of closely spaced peaks) at 0.140 nm and at 0.155 nm, and these account for the pairs of spots in the diffraction pattern.

With the relatively simple diffraction pattern produced by zinc blende, the broad bandwidth and low temporal coherence are only a minor disadvantage. It is possible to identify which spot corresponds to which wavelength quite easily. However, with a more complicated structure it would be far more difficult to untangle which wavelength produced which spot. We would need illumination with higher temporal coherence.

Fortunately filters are available which improve temporal coherence. By passing the X-ray beam through a thin sheet of foil made of a suitable metal, much of the shorter wavelength radiation can be absorbed, so that a far larger proportion of the energy is concentrated in the peaks at 0.155 nm (Figure 49(b)). The effect that this has is evident by comparing the diffraction pattern obtained with the filtered radiation, shown in Figure 50, with the pattern obtained with unfiltered radiation, shown in Figure 48. The filtered, high-temporal-coherence illumination produces a simpler pattern that is easier to interpret. In particular, there is now no danger of associating the wrong wavelength (and hence the wrong spatial frequency) with a spot.

Consider now the spatial coherence of the X-ray illumination. The spots in the diffraction pattern in Figure 50 appear almost point-like, but on the negative from the X-ray camera, they have a diameter of about a millimetre. This size is related not only to the angular size of the source, but also to the diameter of the X-ray beam.

When discussing light, I completely ignored the diameter of the illuminating beam. The reason for this is that we can use lenses to bring extended beams of light to a focus. Thus we can use lenses to expand a laser beam to a diameter of a centimetre, say, use the beam to illuminate a transparency, and then use another lens to produce a diffraction pattern in its back focal plane. The sharpness of the diffraction pattern produced in this way is essentially

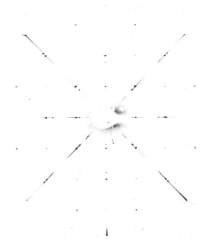

Figure 48 X-ray diffraction pattern of a zinc blende crystal, obtained with temporally incoherent (i.e. unfiltered) X-rays.

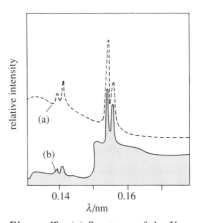

Figure 49 (a) Spectrum of the X-ray beam used to produce Figure 48. (b) Spectrum of the filtered X-ray beam used to produce Figure 50.

Figure 50 X-ray diffraction pattern of a zinc blende crystal, obtained with temporally coherent (i.e. filtered) X-rays.

independent of the diameter of the expanded beam. But the non-existence of X-ray lenses means that an expanded beam cannot be focused, and this is the reason that it is important to control the diameter of an X-ray beam, as well as to control the angular size of the source.

X-rays are produced by accelerating electrons from a heated filament so that they strike a metal target. The **X-ray tube** is designed so that the electrons strike a small area of the metal; this area is the source of the X-rays, and typically it appears to be effectively square, about 1 mm × 1 mm. However, X-rays will diverge from this small source, and the beam is therefore collimated to restrict its diameter. The collimating tube typically has a diameter of about 0.5–1 mm, and with this arrangement, the beam illuminating the crystal is about 1 mm in diameter. For some diffraction studies, a more spatially coherent beam with a smaller diameter is required. X-ray tubes are available with source sizes of less than 0.1 mm, and smaller diameter collimators can also be used. However, both of these changes reduce the intensity of the X-rays reaching the specimen, and they therefore necessitate greater exposure times.

7.1 Summary of Section 7

1 The arrangement of atoms in crystals can be determined by illuminating them with X-rays and making measurements on the diffraction pattern that is produced.

2 The X-ray illumination used for crystallographic measurements needs to be both spatially and temporally coherent.

3 A metal foil filter is used to improve the temporal coherence of the X-ray illumination by restricting the range of wavelengths present.

4 A metal tube collimator is used to improve the spatial coherence of the illumination by restricting the range of inclinations. It also restricts the diameter of the X-ray beam.

SAQ 14 Figure 51 is a schematic representation of the central region of an X-ray diffraction pattern.

(a) Suggest two modifications that need to be made to the equipment to produce an improved diffraction pattern.

(b) State how each of these modifications would affect the diffraction pattern.

(c) State whether the modifications affect the spatial or temporal coherence of the X-ray beam.

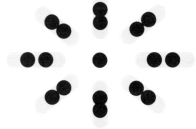

Figure 51 For use with SAQ 14.

ITQ answers and comments

ITQ 1 (a) You should have observed a set of equally spaced spots, distributed along a horizontal line, and this is similar to the pattern that your laser would produce. If the room was dark enough, you may have seen seven spots, but if you were unable to darken the room completely, you probably saw only five, or possibly only three.

(b) I measured that the separations between the central (zero frequency) spot and the spots on either side of it (corresponding to $q = \pm 300$ cycles mm^{-1}) were $s = 18.5$ mm, and the spacing between lens E and the screen was $f = 105$ mm. Using equation 2, the diffraction angle θ is given by

$$\tan\theta = s/f = 18.5/105, \text{ or } \theta = 10.0 \text{ degrees.}$$

Then using equation 1, $\sin\theta = q\lambda$, we obtain

$$\begin{aligned}\lambda &= \sin\theta/q = \sin 10.0/(300 \text{ lines mm}^{-1})\\ &= 5.8 \times 10^{-4} \text{ mm}\\ &= 5.8 \times 10^{-7} \text{ m} = 580 \text{ nm.}\end{aligned}$$

Now since θ is small, I could have used the simpler equation $s = fq\lambda$ (equation 3) with an error of only $1\tfrac{1}{2}\%$ (Table 1). This leads to the result

$$\lambda = s/fq = 18.5 \text{ mm}/(105 \text{ mm} \times 300 \text{ cycles mm}^{-1})$$
$$= 590 \text{ nm,}$$

which is indeed $1\tfrac{1}{2}\%$ larger than the first result. Note that it is the first result (580 nm) that is more accurate, but for our purposes the difference is not significant.

These values for the wavelength correspond to the yellow–green region of the spectrum, which is the colour transmitted by the filter used in the experiment.

The laser-like diffraction pattern, with a spot spacing consistent with the known values of the wavelength of yellow–green light and the grating spatial frequency, indicates that the pseudo-laser light is diffracted in a similar way to a laser beam.

ITQ 2 The only effect of increasing the beam diameter is to make the diffraction pattern brighter. The large-diameter plane wave beams are focused to the same points in the diffraction pattern as the smaller diameter beams, as shown in Figure 52. However, because there is more light energy in the broader beams, they produce a brighter diffraction pattern. Since the larger diameter beam produces a diffraction pattern that is easier to see, I suggest that you do the remaining experiments in this unit *without* the restricting aperture on lens D_2.

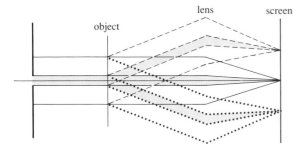

Figure 52 The 2 mm diameter beams (heavy shading) and the larger diameter beams are focused to the same points in the diffraction pattern.

ITQ 3 (a) You should not have observed any overall shift of the diffraction pattern when the yellow–green filter was removed; the pattern remains symmetrical about the central spot. In addition, the yellow–green parts of the pattern are in the same places as they were before removal of the filter.

(b) When a parallel monochromatic beam of light illuminates the 300 lines per millimetre grating, the diffraction spots appear at distances $s = fq\lambda$ (equation 3) from the central spot, *assuming that the diffraction angle θ is small*. In this expression, f is the focal length of lens E that forms the diffraction pattern, and q takes the values 300 cycles mm^{-1}, 600 cycles mm^{-1}, 900 cycles mm^{-1}, etc., which are the spatial frequency components of the grating transmittance. Clearly these distances are proportional to the wavelength λ of the illumination. When a wide range of wavelengths is used, as is the case when the filter is removed, each wavelength component forms its own diffraction pattern, and the scale of the pattern is proportional to the wavelength. The wavelength of red light is longer than that of blue light, $\lambda_\text{red} > \lambda_\text{blue}$, and so the red parts of the diffraction pattern will be further from the centre than the corresponding blue parts.

(c) I estimated the wavelengths by measuring the distances from the central spot of the red and blue ends of the 300 cycles mm^{-1} part of the diffraction pattern. These distances were about 22 mm and 13.5 mm. Using the expression $s = fq\lambda$ again (as in ITQ 1), with $s = 22$ mm for red light, with my measured value for the focal length of lens E, $f = 105$ mm, and with $q = 300$ cycles mm^{-1}, I obtained a value $\lambda_\text{r} = 700$ nm. For blue light, $s = 13.5$ mm and so $\lambda_\text{b} = 430$ nm.

(d) The central spot of the diffraction pattern is the zero spatial frequency component, i.e. $q = 0$. All zero frequency spots appear at the same place, *irrespective of λ*, since $fq\lambda = 0$ when $q = 0$ for *all* values of λ. The combination of all wavelengths produces the white central spot.

ITQ 4 The spectrum of the light from the top of the laser tube is different from both the spectrum of the monochromatic laser beam and the continuous spectrum of the tungsten lamp. Figure 53 shows a somewhat stylized sketch of the spectrum that I observed. There are contributions from wavelengths throughout the visible spectrum, from blue to red, so in this respect the spectrum is similar to that from the tungsten lamp. However, the spectrum consists of over 30 well-defined wavelength 'lines' and not the continuous range of wavelengths emitted by the tungsten lamp, so in this respect the spectrum is similar to that produced by many lasers with different wavelengths.

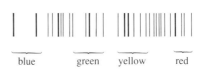

Figure 53 Schematic diagram of the spectrum from the top of the laser tube. All of the lines correspond to $q = 300$ cycles mm^{-1}, but each corresponds to a different value of the wavelength.

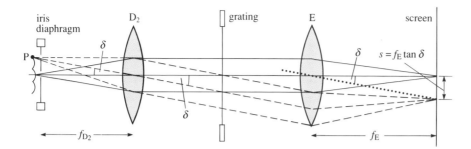

Figure 54 Answer to ITQ 5.

ITQ 5 (a) The path taken by light from point P as it travels to the back focal plane is shown by the broken lines in Figure 54. When drawing this path, I went through the following steps/thought processes, which were introduced in Units 1 & 2:

(i) I drew a beam diverging from P and passing through the same region of lens D_2 as the beam in Figure 10;

(ii) P is in front focal plane of lens D_2, so the beam must be parallel after passing through the lens;

(iii) a ray through the centre of a lens is undeviated, so the parallel beam must be in the direction of the ray from P that passes through the centre of lens D_2;

(iv) the zero-frequency component produced by the grating will travel in the same direction as the incident illumination — it isn't deviated;

(v) a parallel beam is focused by lens E to a point in its back focal plane, and the position of this point is found by drawing a parallel ray that passes undeviated through the centre of the lens (shown as a dotted line in Figure 54).

(b) As Figure 54 shows, the beam illuminating the grating is inclined at the angle δ to the optical axis, and the zero frequency spot is at distance $f_E \tan \delta$ from the optical axis.

ITQ 6 You should have observed that each spot in the diffraction pattern became an image of the filament when the aperture was opened. In the previous experiment you saw that different points on the filament produced displaced diffraction patterns — now with the aperture open you are seeing the diffraction patterns produced by all points on the filament at the same time.

ITQ 7 The diffraction pattern that I observed when the grating lines were vertical was quite complicated, but it was possible to see that it was made up of equally spaced images of the filament. The central image (the zero spatial frequency component) was white, but the other images were dispersed into 'rainbows'. The dispersion of the 600 cycles mm^{-1} images was twice as large as the dispersion of the 300 cycles mm^{-1} images.

When the grating lines were horizontal, the separate effects of source size and bandwidth were much easier to see, because the spreading of the diffraction pattern due to source size was perpendicular to the spreading due to the range of wavelengths. Figure 55 schematically summarizes these observations.

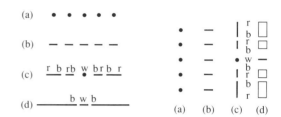

Figure 55 Diffraction patterns produced by a grating when the grating lines are vertical (left) and horizontal (right). In each case, the four patterns shown are produced when the grating is illuminated by (a) monochromatic point source; (b) monochromatic source, extended in horizontal direction; (c) white light from point source; (d) white light source, extended in horizontal direction. The blue and red ends of the white light diffraction patterns are denoted by b and r, respectively, and the white 'straight-through' components by w.

ITQ 8 (a) The graphs illustrating the convolution operation are shown in Figure 56. The transform of the two plane waves that make up the illumination is two delta functions; since the inclinations of the two waves differ by a factor of two, the displacements of the delta functions from the origin also differ by a factor of two. The Fourier transform of the output field is obtained by dealing the transform of the sinusoidal transmittance (three delta functions) to these two delta functions.

(b) The back focal plane pattern due to the illumination alone is shown at the bottom left of Figure 56. Note that the same distance scale is used in this sketch and in the sketch graph above it. The sketch at the bottom right shows the diffraction pattern with the grating in position, and the broken lines show how the spots in the diffraction pattern are related to the delta functions in the graph above.

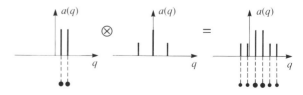

Figure 56 Answer to ITQ 8.

ITQ 9 (a) Spontaneous emission of a photon is associated with a transition from a higher energy state to a lower energy state, so transitions B and C could be spontaneous emission processes. When a photon is absorbed, the energy of the atom increases, and this is what happens in transition A. Note that the transitions shown in Frame 5 are not necessarily accompanied by emission or absorption of photons. The energy could be supplied by an electrical discharge, or energy could be lost in a collision with another atom.

(b) The photon energy hf is equal to the energy difference ΔE between the two states involved in the transition. The frequency is therefore $f = \Delta E/h$, and the wavelength is $\lambda = c/f$, where c is the speed of light. The values for the three transitions are tabulated below:

Transition	A	B	C
$hf/10^{-19}\,\mathrm{J}\,(=\Delta E)$	2	8	4
$f/10^{14}\,\mathrm{Hz}\,(=\Delta E/h)$	3	12	6
$\lambda/\mathrm{nm}\,(=c/f)$	1000	250	500

Note that only transition C is associated with a photon in the visible region of the spectrum. Transition A corresponds to an infrared photon and B to an ultraviolet photon.

ITQ 10 (a) $y_1 = x_1^2 = 9$, and $y_2 = x_2^2 = 1$.

(b) $x_r = x_1 + x_2 = 3 + 1 = 4$, and so $y_r = x_r^2 = 16$. The value of $y_1 + y_2$ is 10. So although $x_r = x_1 + x_2$, it is *not* true that $y_r = y_1 + y_2$. Because of the non-linear relationship between y and x, individual values of y cannot be added to determine the resultant y_r. In fact

$$\begin{aligned} y_r = x_r^2 &= (x_1 + x_2)^2 \\ &= x_1^2 + x_2^2 + 2x_1 x_2 \\ &= y_1 + y_2 + 2x_1 x_2. \end{aligned}$$

The importance of this last term to the problem of adding intensities will soon become apparent.

ITQ 11 For the situation shown in Figure 18(a), $\psi_1 = \psi_2$, and so $I_1 = I_2$. Using equation 10, which is true for all waves,

$$\begin{aligned} I_r &= I_1 + I_2 + 2\langle \psi_1 \psi_2 \rangle \\ &= I_1 + I_1 + 2\langle \psi_1^2 \rangle \\ &= I_1 + I_1 + 2I_1 = 4I_1. \end{aligned}$$

So the resultant intensity is twice the sum of the individual intensities.

ITQ 12 Delta functions have infinitesimal width, so the implicit assumption is that the slits are extremely narrow. For more realistic slits, each delta-function transmittance should be replaced by a narrow top-hat function, as shown in Figure 57. This pair of top-hat functions can be regarded as the *convolution* of a top-hat function with a pair of delta functions. The Fourier transform is therefore the *product* of the Fourier transform of a top-hat — a sinc function — and the Fourier transform of the pair of delta functions — a cosine function. This product is shown on the right of Figure 57.

So, in practice, the diffraction pattern of a double slit is not a set of cosinusoidal fringes, all with the same intensity. The intensity of the fringes decreases away from the centre of the pattern, the fringes disappear, and then reappear again. *However, we will ignore this complication in most of the discussion that follows. We will assume that we are dealing with slits that have infinitesimal width, so that the cosinusoidal fringes in the diffraction pattern all have the same intensity.*

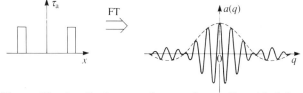

Figure 57 Amplitude transmittance of two slits with finite widths, and the corresponding Fourier transform.

ITQ 13 According to equation 18, the bandwidth $\Delta\lambda$ is given by $\Delta\lambda \approx \lambda_{\mathrm{av}}/n$, where n is the number of fringes observed on either side of the central fringe. Thus

$$\Delta\lambda \approx 530\,\mathrm{nm}/8 = 66\,\mathrm{nm},$$

and this means that the wavelengths in the green light are roughly in the range 500 nm to 560 nm.

ITQ 14 (a) For $\theta = 0$ (Figure 32(a)), the light from the two slits will travel the same distance to the diffraction pattern, so superposition will lead to complete constructive interference — a bright fringe. The angle shown in (b) also corresponds to constructive interference: the light waves are in phase because light from slit S_2 travels exactly one wavelength further than light from S_1. The diagram in Figure 32(c) shows that light from S_2 has to travel an extra distance of $3\lambda/2$ compared with light from S_1, so in this case the waves are in antiphase, and so the interference is destructive, resulting in a dark fringe. Finally, part (d) of Figure 32 shows light from S_2 travelling about 1.7λ further: this is intermediate between the conditions for complete constructive or complete destructive superposition.

(b) For complete constructive interference, the difference between the paths taken by light from the two slits must be a whole number n of wavelengths. But by applying trigonometry to Figure 32(b), it can be shown that the extra distance travelled by light from the bottom slit is $d\sin\theta$, where d is the slit separation. The condition for constructive interference is therefore

$$d\sin\theta = n\lambda.$$

This equation may be familiar from previous physics courses.

ITQ 15 (a) For all of the cosinusoidal intensity distributions shown at the top of Figure 38, the minimum intensity $I_{\min}$ is zero. From the definition of modulation,

$$M = \frac{I_{\max} - I_{\min}}{I_{\max} + I_{\min}}, \qquad \text{(Eq. 23)}$$

it immediately follows that $M = 1$.

(b) I used a ruler to measure the relative values of $I_{\max}$ and $I_{\min}$ for the three diffraction patterns at the bottom of Figure 38, and got the following results.

Figure 38(a): $I_{\max} \approx 15\,\mathrm{mm},\ I_{\min} \approx 1\,\mathrm{mm},$

$$M \approx \frac{15-1}{15+1} \approx 0.9.$$

Figure 38(b): $I_{\max} \approx 12\,\mathrm{mm},\ I_{\min} \approx 3\,\mathrm{mm},$

$$M \approx \frac{12-3}{12+3} = 0.6.$$

Figure 38(c): here I is constant, so $I_{max} = I_{min}$ and therefore $M = 0$.

ITQ 16 (a) Assuming that the star acts as a uniform-disc light source, its angular diameter α can be determined from equation 28:

$$w_c = 1.22\lambda_{av}/\alpha.$$

Here w_c is the coherence width of the illumination from the star, and is equal to the mirror separation at which the fringes disappear. Thus

$$\begin{aligned}\alpha = 1.22\lambda_{av}/w_c &= 1.22 \times (570 \times 10^{-9}\,\text{m})/3.07\,\text{m} \\ &= 2.3 \times 10^{-7}\text{ radians,} \\ &= 2.3 \times 10^{-7} \times \frac{360}{2\pi}\text{ degrees} \\ &= 1.3 \times 10^{-5}\text{ degrees} \\ &= 1.3 \times 10^{-5} \times 3\,600\text{ seconds of arc} \\ &= 0.047\text{ seconds of arc.}\end{aligned}$$

(b) For the illumination from the smallest measurable star, the fringes will disappear at the maximum mirror separation (240 inches). The angular diameter of this star will be (121/240) times the diameter of Betelgeuse. This is 1.2×10^{-7} radians, or 0.024 seconds of arc.

Note The value of 0.047 seconds of arc for Betelgeuse is based on the assumption that a star is a *uniform*-disc light source. In fact, stars usually appear darker towards the edges. In 1920, this was only known to be the case for the Sun, but it did mean that there was an extra uncertainty in the value of the diameter to be added to the other experimental uncertainties.

SAQ answers and comments

SAQ 1 (a) The diffraction pattern is produced by an extended source, which is in the form of a line parallel to the lines of the grating. In general, the shape of the zero-frequency 'spot' in a grating diffraction pattern will be an image of the source, and for a monochromatic source all of the other spatial frequency components will have the same form. Since the set of lines in the diffraction pattern are spread out horizontally, the grating lines must be in the vertical direction.

(b) We showed in Section 2.8 that the field in the diffraction plane is $FT(E_{out})$, and is related to the illuminating field E_{in} and the grating transmittance τ_a by the equation

$$FT(E_{out}) = FT(E_{in}) \otimes FT(\tau_a) \qquad \text{(Eq. 5)}$$

Now $FT(\tau_a)$ for the grating is a set of equally spaced spots along the horizontal axis, and $FT(E_{in})$ is essentially the image in the back focal plane of the source, which is a vertical line. Dealing out this vertical line image to each of the spots on the horizontal axis reproduces the diffraction pattern in Figure 15.

(c) (i) Rotating the grating changes τ_a, so that $FT(\tau_a)$ becomes a vertical column of spots, but $FT(E_{in})$ is unchanged. The diffraction pattern will be a set of vertical lines, equally spaced along a vertical line, as shown in Figure 58(a).

(ii) Rotating the source changes E_{in} so that $FT(E_{in})$ becomes a *horizontal* line in the diffraction plane, but $FT(\tau_a)$ is the same as for the original illumination. The pattern will be horizontal lines, equally spaced along a horizontal line, as shown in Figure 58(b).

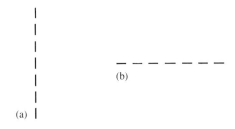

Figure 58 Answers to SAQ 1 (c).

SAQ 2 (a) Patterns A, C and E are formed by monochromatic light. In each of these cases, an image of the source is dealt out to each point of a rectangular array of grid points. The sources for A, C and E are a horizontal line, a disc and two points separated in the horizontal direction, respectively.

(b) Point sources always produce a point of illumination in the diffraction pattern corresponding to the zero spatial frequency component. Patterns B and D fulfil this criterion. For B, the radial lines in the pattern, with length increasing as distance from the centre increases, indicate that the source has a continuous distribution of wavelengths. For D, the source emits just two wavelengths.

(c) Pattern F is not produced by a monochromatic source or by a point source. It is produced by a source with the same bandwidth as produced B and the same shape as produced pattern A. Thus F is equal to B convoluted with the horizontal line that appears in A.

SAQ 3 (a) The lowest spatial frequencies are 120 cycles mm^{-1} in the horizontal direction and 60 cycles mm^{-1} in the vertical direction.

Variation of the transmittance in the horizontal direction produces information along the horizontal axis in the diffraction pattern. The lowest spatial frequency in the horizontal direction gives rise to the horizontal lines on either side of the origin in pattern B, and lines are observed, rather than points, because of the large bandwidth of the source. The ends of these lines closest to the origin correspond to the shortest wavelength (500 nm) in the illumination, and I measured the ends to be about 6 mm from the origin. Making use of the expression $s = fq\lambda$ (Eq. 3),

$$6 \times 10^{-3}\,\text{m} = (100 \times 10^{-3}\,\text{m}) \times q \times (500 \times 10^{-9}\,\text{m}),$$

and hence

$$q = 1.2 \times 10^5 \text{ cycles m}^{-1} = 120 \text{ cycles mm}^{-1}.$$

Similarly, the lowest spatial frequency component in the vertical direction can be estimated from the closest distance of the lines above and below the origin, which I measured as 3 mm. Thus the lowest spatial frequency in the vertical direction is obtained by substituting 3×10^{-3} m instead of 6×10^{-3} m in the equations above, and this leads to a lowest spatial frequency of 60 cycles mm^{-1}.

(b) The ratio of maximum to minimum wavelength in the illumination is equal to the ratio of the distances from the origin of the ends of any line in pattern B. For the longest horizontal line in the pattern, this ratio is about (17 mm)/(13 mm) = 1.3. Since the minimum wavelength is 500 nm, the maximum is 1.3×500 nm = 650 nm.

SAQ 4 (a) The three major processes involving photons are spontaneous emission of a photon (Frame 2 of the AV section), stimulated emission of a photon (Frame 6) and absorption of a photon (Frame 4).

(b) (i) Laser light results from stimulated emission, and this accounts for the fact that vast numbers of photons are emitted with the same frequency and produce a highly directional, monochromatic beam. (ii) Light from a tungsten lamp is produced by spontaneous emission from atoms that have been thermally excited to high energy states. However, all three processes occur in both types of light source. It is the relative probabilities of the processes that account for the differences between the light from a laser and a tungsten lamp.

SAQ 5 The probability of an atom being in a low-energy state is generally greater than the probability of it being in a higher-energy state. This means that more atoms are generally in the ground state than in any other state, and for two states with energies E_1 and E_2, the number of atoms in state E_2 is less than the number in state E_1 if E_2 is greater than E_1. If the opposite is true, that is there are *more* atoms in the *higher*-energy state E_2 than in the lower-energy state E_1, then we say that there is a population inversion.

Population inversion is essential for laser action because photon absorption always competes with stimulated emission in a laser. A photon that has energy corresponding to the energy difference between the two levels that are responsible for laser action can stimulate emission of a photon by causing a transition of an atom from the higher-energy state to the lower-energy state. But it can also be absorbed while exciting an atom from the lower-energy state to the higher state. The relative probabilities of these processes are simply proportional to the numbers of atoms in the two states. For stimulated emission to be more likely than absorption, there must be more atoms in the higher state, i.e. there must be a population inversion for the two states.

SAQ 6 The resultant intensity I_r is given by

$$I_r = I_1 + I_2 + 2\langle \psi_1 \psi_2 \rangle. \qquad (\text{Eq. 10})$$

Since the amplitude of ψ_1 is 1.5 times the amplitude of ψ_2, and since $I = \langle \psi^2 \rangle$ (Eq. 7), it follows that $I_1 = (1.5^2) I_2 = 2.25 I_2$. Thus

$$\begin{aligned} I_r &= 2.25 I_2 + I_2 + 2\langle \psi_1 \psi_2 \rangle \\ &= 3.25 I_2 + 2\langle \psi_1 \psi_2 \rangle. \end{aligned}$$

This is true whatever the phase difference $\Delta\phi$ between the two waves. However, the value of $\langle \psi_1 \psi_2 \rangle$ depends on $\Delta\phi$, and it can be evaluated for the three cases in question by comparison with the equal-amplitude situations shown in Figure 18.

(a) When $\Delta\phi = 0$ (Figure 18(a)), one wave is simply a scaled-up version of the other, i.e. $\psi_1 = 1.5\,\psi_2$. Hence

$$\langle \psi_1 \psi_2 \rangle = \langle 1.5\psi_2 \times \psi_2 \rangle = 1.5 \langle \psi_2^2 \rangle = 1.5 I_2,$$

since, by definition, $I_2 = \langle \psi_2^2 \rangle$. So

$$I_r = 3.25 I_2 + 2 \times 1.5 I_2 = 6.25 I_2.$$

(b) When $\Delta\phi = \pi$ (Figure 18(e)), $\psi_1 = -1.5\,\psi_2$. In this case, $\langle \psi_1 \psi_2 \rangle = -1.5 I_2$, and so $I_r = 0.25 I_2$.

(c) When $\Delta\phi = \pi/2$, $\langle \psi_1 \psi_2 \rangle = 0$, just as it is in the equal amplitude case (Figure 18(c)). Thus $I_r = 3.25 I_2$.

The alternative, and simpler, way to determine the intensities when $\Delta\phi = 0$ or π is to first evaluate ψ_r, and then use the relation $I_r = \langle \psi_r^2 \rangle$ to determine the resultant intensity. Thus

$\Delta\phi = 0$:
$$\begin{aligned} \psi_r &= \psi_1 + \psi_2 \\ &= 1.5\,\psi_2 + \psi_2 = 2.5\,\psi_2; \\ I_r &= \langle \psi_r^2 \rangle \\ &= \langle (2.5\psi_2)^2 \rangle = 6.25 \langle \psi_2^2 \rangle = 6.25 I_2. \end{aligned}$$

$\Delta\phi = \pi$:
$$\begin{aligned} \psi_r &= -1.5\,\psi_2 + \psi_2 = -0.5\,\psi_2; \\ I_r &= \langle (-0.5\psi_2)^2 \rangle = 0.25 I_2. \end{aligned}$$

SAQ 7 Constructive and destructive interference would only be observable in case (c), the region behind the laser-illuminated grating. In this case, the light arriving at any point from each of the slits in the grating originates from a single source and will have the same frequency. Cross-terms, such as $\psi_1 \psi_2$, will not generally average to zero, so constructive and destructive interference will be evident. Interference would *not* be observed in the situations described in (a) and (b). In case (a), the laser beams have different frequencies, so the cross-term will average to zero everywhere, and $I = I_1 + I_2$ as discussed in Section 4.3. In case (b), the sources are independent, so the phase changes in the two wave fields mean that the cross-term $\psi_1 \psi_2$ time-averages to zero, even though the waves have the same average frequency. Thus, in this case too, the total intensity is simply the sum of the intensities of the individual waves.

SAQ 8 (a) In Experiment 4 (Section 2.5.1), you should have observed that the tungsten lamp produced a continuous spectrum, covering the visible range from blue to red light. ITQ 3(c) asked you to calculate the wavelengths of the blue and red ends of the spectrum from measurements on the grating diffraction pattern, and I obtained the values $\lambda_b = 430$ nm and $\lambda_r = 700$ nm (see answer to ITQ 3(c)). Hence the bandwidth $\Delta\lambda$ is about $(700 - 430)$ nm = 270 nm.

(Note that this probably overestimates the bandwidth because I measured the extremes of the visible spectrum, rather than the positions where the intensity fell to half of its maximum value.)

(b) From equation 17, the frequency bandwidth $\Delta f \approx (f_{\text{av}}/\lambda_{\text{av}})\Delta\lambda$, and from equation 16, $f_{\text{av}} \approx c/\lambda_{\text{av}}$. Combining these two equations to eliminate f_{av} leads to

$$\Delta f \approx (c/\lambda_{\text{av}}^2)\Delta\lambda.$$

Substituting for c, $\Delta\lambda$ and λ_{av} ($= \tfrac{1}{2}(430 + 700)$ nm),

$$\begin{aligned} \Delta f &\approx \frac{3 \times 10^8 \text{ m s}^{-1}}{(565 \times 10^{-9} \text{ m})^2} \times 270 \times 10^{-9} \text{ m} \\ &= 2.5 \times 10^{14} \text{ Hz}. \end{aligned}$$

Then, using equation 19,

$$\text{coherence time } t_c = 1/\Delta f = 4 \times 10^{-15} \text{ s},$$

and using equation 20,

$$\begin{aligned} \text{coherence length } l_c &= c t_c \\ &= 3 \times 10^8 \text{ m s}^{-1} \times 4 \times 10^{-15} \text{ s} \\ &= 1.2 \times 10^{-6} \text{ m}. \end{aligned}$$

SAQ 9 (a) Two bright fringes can be detected on either side of the central fringe. As was stated in Section 5.3.1, the number n of bright fringes on each side of the central fringe is given by

$$n \approx \lambda_{\text{av}}/\Delta\lambda. \qquad (\text{Eq. 18})$$

For the tungsten lamp in the Home Kit, $\lambda_{\text{av}} \approx 565$ nm and $\Delta\lambda \approx 270$ nm (see answer to SAQ 8). Thus

$$n \approx 565/270 = 2 \text{ fringes}.$$

So, 2 fringes will be detected on each side of the centre. Note that if these fringes were observed by eye, a continuous colour variation would be apparent for each bright fringe, with blue closest to the centre of the pattern and red furthest from the centre. With a photometer, though, only the smooth variation of total intensity would be detected.

(b) Interference fringes would *not* be observable. In SAQ 8, the coherence length of the tungsten lamp illumination was estimated to be about 10^{-6} m. This is very much smaller than the distance between the mirrors in Figure 22(a), so there will be no phase coherence between the two beams — and therefore no interference fringes — where they overlap in region P.

(c) Interference fringes will be observed in region P when a laser with coherence length 0.5 m is used. With this source, the coherence length is much larger than the difference between the paths travelled by the two beams. Thus there will still be phase coherence between the two beams, and so interference fringes will be observed.

SAQ 10 (a) The coherence length is 0.3 m. Fringes disappear when the beam travelling via mirror M_1 has an extra path length of 2×0.15 m (i.e. there and back). Disappearance of the fringes indicates lack of phase coherence between the two beams, and the minimum path-length difference for which this is true is defined as the coherence length.

(b) Coherence time τ_c = (coherence length l_c)/(speed of light c) = $(0.3 \text{ m})/(3 \times 10^8 \text{ m s}^{-1})$ = 10^{-9} s.

So for this laser, there is no phase coherence between the light emitted at one instant and the light emitted 10^{-9} s later.

(c) Bandwidth $\Delta f = \tau_c^{-1} = 10^9 \text{ s}^{-1} = 10^9$ Hz. To express the bandwidth as a wavelength spread $\Delta \lambda$, we use the expressions $\Delta \lambda / \lambda_{av} \approx \Delta f / f_{av}$, and $f_{av} \lambda_{av} \approx c$. Thus

$$\Delta \lambda \approx \Delta f \times \frac{\lambda_{av}^2}{c} = 10^9 \text{ Hz} \times \frac{(633 \times 10^{-9} \text{ m})^2}{3 \times 10^8 \text{ m s}^{-1}}$$
$$= 1.3 \times 10^{-12} \text{ m} = 1.3 \times 10^{-3} \text{ nm}.$$

This is very small compared with the wavelength of the laser, 633 nm. Contrast these results with the coherence length, the coherence time and the bandwidth of a tungsten lamp calculated in SAQ 8. The laser has a much greater temporal coherence than the tungsten lamp — by a factor of about 10^5!

SAQ 11 (a) After adjusting the diameter of the iris diaphragm so that the filament images were just touching, I measured the diameter D as 9 mm. I assumed that the focal length of lens D_2 was the distance between the iris and D_2, since this distance was adjusted so that the lens produced a parallel beam; this distance was 51 mm. Then using equation 22, the range of inclinations of the plane waves reaching the grating was

$$\alpha = D/f = 9 \text{ mm}/51 \text{ mm} = 0.18 \text{ radians}.$$

(b) For a line source, the coherence width w_c is given by equation 27, $w_c = \lambda_{av}/\alpha$. Hence

$$w_c = (580 \times 10^{-9} \text{ m}) / (0.18 \text{ radians}) = 3.2 \times 10^{-6} \text{ m}.$$

(c) The spatial period of the grating is the reciprocal of the spatial frequency of 300 cycles mm^{-1}, so the spatial period is $(300 \text{ mm}^{-1})^{-1}$, or 3.3×10^{-3} mm, or 3.3×10^{-6} m. This is approximately equal to the coherence width calculated in part (b).

The filament images touch when their angular length α is equal to the spacing θ of the frequency components in the diffraction pattern. Now, $\sin \theta = q \lambda$, or $\theta = q \lambda$ since θ is small, so the condition for the images to touch is $\alpha = \theta$, or $\alpha = q \lambda$, or $q^{-1} = \lambda / \alpha$. Since q^{-1} is the spatial period of the grating, and λ/α is the coherence width of the illumination, these two quantities should be equal when the filament images just touch.

An alternative approach is to say that if the filament images touch, the effective 'modulation' of the diffraction pattern is zero, and this happens when the coherence width is equal to the spacing of the grating lines.

SAQ 12 (a) For a circular source, the coherence width w_c is given by $w_c = 1.22 \lambda_{av}/\alpha$ (equation 28), and α is given by equation 21, $\alpha = D/r$. Thus

$$\alpha = D/r = (50 \times 10^{-3} \text{ m})/10 \text{ m} = 5 \times 10^{-3} \text{ radians},$$
and
$$w_c = 1.22 \lambda_{av}/\alpha = 1.22 \times (650 \times 10^{-9} \text{ m})/5 \times 10^{-3}$$
$$= 1.6 \times 10^{-4} \text{ m}.$$

(I have assumed that the average wavelength of the red light is 650 nm.)

(b) The slit separation is 10^{-4} m, so (slit separation)/(coherence width) = 0.63. From Figure 46, the modulation is about 0.4 at this value of the ratio of slit separation to coherence width.

(c) The modulation depends on the coherence width, which in turn depends on the angular diameter of the source. To produce fringes with the same modulation, the angular size, $\alpha = D/r$, must be the same. Since the distance r is reduced by a factor of 10, the source diameter must be reduced by the same factor, so the aperture diameter must be 50 mm/10 = 5 mm.

SAQ 13 (a) The fringes produced by the two stars will cancel out when the intensity maxima produced by one star coincide with the intensity minima due to the other star. This will happen when the angular shift between the two sets of fringes (which is equal to the angular separation α of the two stars) is half of the spacing of the peaks of the cosinusoidal pattern. Since the peak spacing is given by $\sin \theta = \lambda/d$, or $\theta = \lambda/d$ assuming that θ is small, the condition for fringe disappearance is

$$\alpha = \tfrac{1}{2}\theta = \lambda_{av}/2d.$$

(b) Since $d = 4.5$ m, and $\lambda_{av} = 500$ nm,

$$\alpha = (500 \times 10^{-9} \text{ m})/(2 \times 4.5 \text{ m}) = 5.6 \times 10^{-8} \text{ radians}.$$

(c) Since α = separation/distance,

$$\text{separation} = 5.6 \times 10^{-8} \text{ radians} \times 3 \times 10^{19} \text{ m}$$
$$= 2 \times 10^{12} \text{ m}.$$

SAQ 14 The size of the 'spots' in the diffraction pattern indicate that a large diameter beam was used to produce Figure 51, and the double spots and the radial streaks indicate that an unfiltered beam was used.

(a) Two modifications that would sharpen the diffraction pattern are a narrow collimator to reduce the diameter of the beam striking the crystal, and a filter to remove one of the peaks in the spectrum and to remove much of the continuous background radiation (see Figure 49).

(b) The collimator will reduce the diameter of the spots, and the filter will replace each pair of spots and the associated streak with a single spot.

(c) The collimator improves the spatial coherence and the filter improves the temporal coherence.

Index of important terms and concepts

absorption (of photon) 20
bandwidth 30
beam-splitter 28
coherence length 38
coherence time 37
coherence width 45
constructive interference 25
convolution theorem 14
correlation (between waves) 28
cross-term $\psi_1\psi_2$ 24
destructive interference 25
diaphragm 8

energy state 19
excited state 19
Fourier transform spectrometry 39
ground state 19
intensity 23
laser process 20–22
Michelson interferometer 39
Michelson stellar interferometer 47
modulation 43
photon energy 19
population inversion 21
principle of superposition 23

solar spectrum 20
spatial coherence 41
spectral bandwidth 7
spectrum 7
spontaneous emission 19
stimulated emission 20
superposition 23
temporal coherence 30
visibility 43
X-ray diffraction 50
X-ray tube 52
X-rays 50